Workbook

Engineering Fundamentals

Design, Principles, and Careers

by

Ryan A. Brown

Joshua W. Brown

Michael Berkeihiser

Publisher
The Goodheart-Willcox Company, Inc.
Tinley Park, IL
www.g-w.com

Manufactured in the United States of America.

ISBN 978-1-61960-225-0

2 3 4 5 6 7 8 9 – 14 – 18 17 16 15

Cover credit: Torsten Lorenz/Shutterstock.com

Introduction

This workbook is designed for use with the text ***Engineering Fundamentals***. The chapters in the workbook correspond to those in the text and should be completed after reading the appropriate text chapter.

Each chapter of the workbook contains reviews of the textbook chapters to enhance your understanding of textbook content. The various types of questions include matching, true or false, multiple choice, fill-in-the-blank, and short answer.

The workbook chapters also contain activities related to textbook chapter content. The activities range from chapter content reinforcement to real-world application, including design projects related to different engineering disciplines. It is important in these activities to understand any safety procedures set forth by your teacher.

Contents

Name ______________________________ Date ____________ Class __________

Introduction Activity

Safety

Safety is the single most important consideration when working in any laboratory setting. If you fail to follow the safety rules described by your teacher, you can cause serious bodily injuries to yourself or others.

Write down the safety rules your instructor has developed for your specific lab situation.

Chapter 1 Review

What Is Engineering?

Name ______________________________ Date ____________ Class __________

1. What is an *engineer*?

____

____

2. What is a specific set of steps that begins with a problem and ends with a solution?

____

____

3. Describe the difference between specifications and constraints.

____

____

____

Matching

_______ 4. An understanding of data and charts.

_______ 5. An understanding of concepts such as statics and circuit theory.

_______ 6. An understanding of the use of tools.

A. Mathematical knowledge

B. Technical knowledge

C. Scientific knowledge

7. List three skills that are important for engineers to possess.

____

____

____

____

8. Describe the role of an engineering technician.

____

____

____

_______ 9. The _____ engineering discipline is concerned with the design and building of large-scale construction projects.

A. civil

B. mechanical

C. aerospace

D. geomatics

_______ 10. The _____ engineering discipline is concerned with the design of artificial organs and prosthetics.

A. civil

B. bioengineering

C. computer

D. chemical

_______ 11. The _____ engineering discipline is concerned with the design and building spacecraft.

A. civil

B. mechanical

C. aerospace

D. manufacturing

_______________ 12. True or False? The materials that have been available throughout history have impacted the work of engineers.

13. During which time period was the steam engine created?

____

____

14. Name one branch of engineering that had its start in the twentieth century.

____

____

_______________ 15. True or False? Engineering is a profession that is coming to an end because all possible products have been designed.

Name ______________________________ Date ______________ Class ____________

Activity 1-1

Engineering Requirements

The role of an engineer revolves around optimizing a solution that balances the trade-offs between specifications and constraints. In this activity, you will make trade-offs between specifications and constraints for an existing product.

Objective

After completing this activity, you will be able to:

- Provide examples of specifications, constraints, and trade-offs.

Materials

Pencil

Activity 1-1 Worksheet

Activity

In this activity, you will:

1. Select an engineered product, such as a bridge, computer, or electrical device.
2. Imagine you are the engineer who is asked to design the selected product. Use the Activity 1-1 Worksheet to list the specifications (requirements) and the constraints (limitations) that you must consider when designing the product.
3. On the Activity 1-1 Worksheet, make a list of trade-offs that you would be willing to make as the designer of the product.
4. Present your findings to the rest of the class.

Reflective Questions

1. How are specifications and constraints different from each other?

2. Why is it important for an engineer to make trade-offs between specifications and constraints?

Name ______________________________

Activity 1-1 Worksheet

Specifications	Constraints

Trade-Offs

Notes

Name ______________________ Date __________ Class __________

Activity 1-2

Knowledge Within Engineering Disciplines

All engineers use a combination of mathematical, scientific, and technical knowledge. In this activity, you will determine examples of the types of knowledge used in an engineering discipline.

Objective

After completing this activity, you will be able to:

- Describe different examples of mathematical, scientific, and technical knowledge.

Materials

Pencil

Activity 1-2 Worksheet

Activity

In this activity, you will:

1. Select an engineering discipline, such as civil engineering, mechanical engineering, computer engineering, or aerospace engineering.
2. Conduct research on the engineering discipline to gain an understanding of it.
3. Use the included Activity 1-2 Worksheet to generate a list of the examples of the types of knowledge that an engineer in the selected discipline would use or need to understand.
4. Present your findings to the rest of the class.

Reflective Questions

1. How are the three types of knowledge different from one another?

____

____

____

2. How do you believe engineers develop knowledge in each of the three areas?

____

____

____

Activity 1-2 Worksheet

Mathematical Knowledge	Scientific Knowledge	Technical Knowledge

Name ______________________________ Date ____________ Class ____________

Activity 1-3

Engineering Advancements

The history of technology is very closely associated with advancements that have been made in engineering. In this activity, you will select a technological device and trace its history and predict its future.

Objective

After completing this activity, you will be able to:

- Discuss technological and engineering development and change.

Materials

Books or eBooks about technological devices, products, or systems

Internet access

Presentation software

Activity

In this activity, you will:

1. Select a technological device, product, or system, such as an airplane, road, computer, or cell phone.
2. Conduct research on the device, product, or system you selected. Find answers to the following questions:
   A. When was the device, product, or system first designed/developed?
   B. Who first designed/developed it?
   C. What types of engineers are (or may be) involved in the design and operation of the product?
   D. What products or devices came before that may have been used in the design/development?
   E. What knowledge and materials had to exist before the product could be designed/developed?
   F. What problem does this product solve?
   G. What new products could this product lead to in the future?
3. Create a presentation using presentation software to present your findings.
4. Present the development of your product to your class.

Reflective Questions

1. What did you learn about the development of technology and the role of engineers?

2. How do you believe the growth of knowledge and understanding of engineering has impacted the products that we have today?

Chapter 2 Review

Engineering Design

Name ______________________________ Date ____________ Class ___________

_______ 1. _____ is the creative application of technology to design a system, product, or process to solve a given problem or meet a given need.

A. Artistic design

B. Engineering design

C. Troubleshooting

D. Market research

____________________ 2. True or False? Engineering design is important to all engineering disciplines. Regardless of individual area, all engineers can use the design process to solve complex problems.

3. List the steps of the design process in their proper order.

____

____

____

____

____

____________________ 4. True or False? The engineering design process is a fixed series of six steps used by all engineers.

_______ 5. The _____ outlines the problem in clear terms.

A. idea generation

B. problem solving

C. problem statement

D. final solution

_______________ 6. _____ are the limitations of a design.

_______________ 7. At the _____ stage, there are no wrong answers. It is important to come up with as many solutions as possible and keep good records of each solution.

_______________ 8. At the _____ stage, all ideas are evaluated and the best one is selected.

Matching

_____ 9. Converts drawings to 3-D shapes.

_____ 10. Software that can create line drawings and 3-D models.

_____ 11. Full-color drawings.

A. Renderings

B. Computer numerical control

C. Computer-aided drafting

_______________ 12. At the _____ stage, computer simulations and prototypes are used to evaluate solutions.

_______________ 13. At the _____ stage, a decision has been made for a solution and it must be communicated in great detail to the people who will make that product and buy the product.

_______________ 14. The _____ step occurs after a final solution has been made and possibly even been produced.

_______________ 15. _____ are written records kept by engineers detailing everything associated with the design process.

Name ______________________________ Date ____________ Class ___________

Activity 2-1

Problem Definition

Problem definition is often considered to be the most important step in the design process. It is critical that the correct problem be identified as outlined in your text. Constraints are the specific conditions that solutions must meet in order to solve the problem. Constraints and criteria could include things like size, weight, color, and cost. In this activity, you will identify problems as well as criteria and constraints.

Objectives

After completing this activity, you will be able to:

- Identify a design problem.
- Write an effective problem statement.
- Identify the constraints and criteria for a given problem.

Materials

Pencil

Activity

In this activity, you will:

1. Identify a problem you see in the products you use in your daily life.
2. Think about how you could redesign an existing product to meet that need.
3. Decide whether you would need to design a completely new product.
4. Identify the constraints and criteria for the product.

Reflective Questions

1. Where did the idea for this problem originate? Was it your idea, or did it come from someone else?

2. In the space below, write everything you know about the problem. If the idea came from someone else, you may need to speak with that person to fully understand the problem.

3. Think about the need you are addressing in this design problem and write a problem statement on the lines below. Make sure your statement is specific enough to fully describe the problem, but not so specific that it limits the creativity of your design.

4. Determine the design constraints and criteria for your problem and write them in the space below.

Name ______________________________ Date ____________ Class __________

Activity 2-2

Idea Generation—Brainstorming

The goal of the idea generation step is to come up with as many design solutions as possible and to keep a good record of them for further evaluation later. There is no evaluation at this point, and there are no wrong answers. One of the most effective methods of generating ideas is brainstorming. In this activity, you will use brainstorming to get as many ideas on paper as possible.

Objective

After completing this activity, you will be able to:

- Generate a wide variety of creative solutions to a design problem.

Materials

Pencil

Activity 2-2 Worksheet

Activity

In this activity, you will:

1. Bring together a group of classmates for a brainstorming session, including some who are not involved in your design project.
2. Explain the problem, the constraints, and the findings of your research.
3. Using Activity 2-2 Worksheet, write down every idea that comes up in your session, regardless of what you think of the idea.
4. Draw sketches as necessary.
5. Keep accurate records of all ideas so you can evaluate them at a later time.

Reflective Questions

1. Why is it preferable to write down as many ideas as possible?

2. Did it help to get ideas from your classmates who are not involved in your project? Why or why not?

Name ____

Activity 2-2 Worksheet

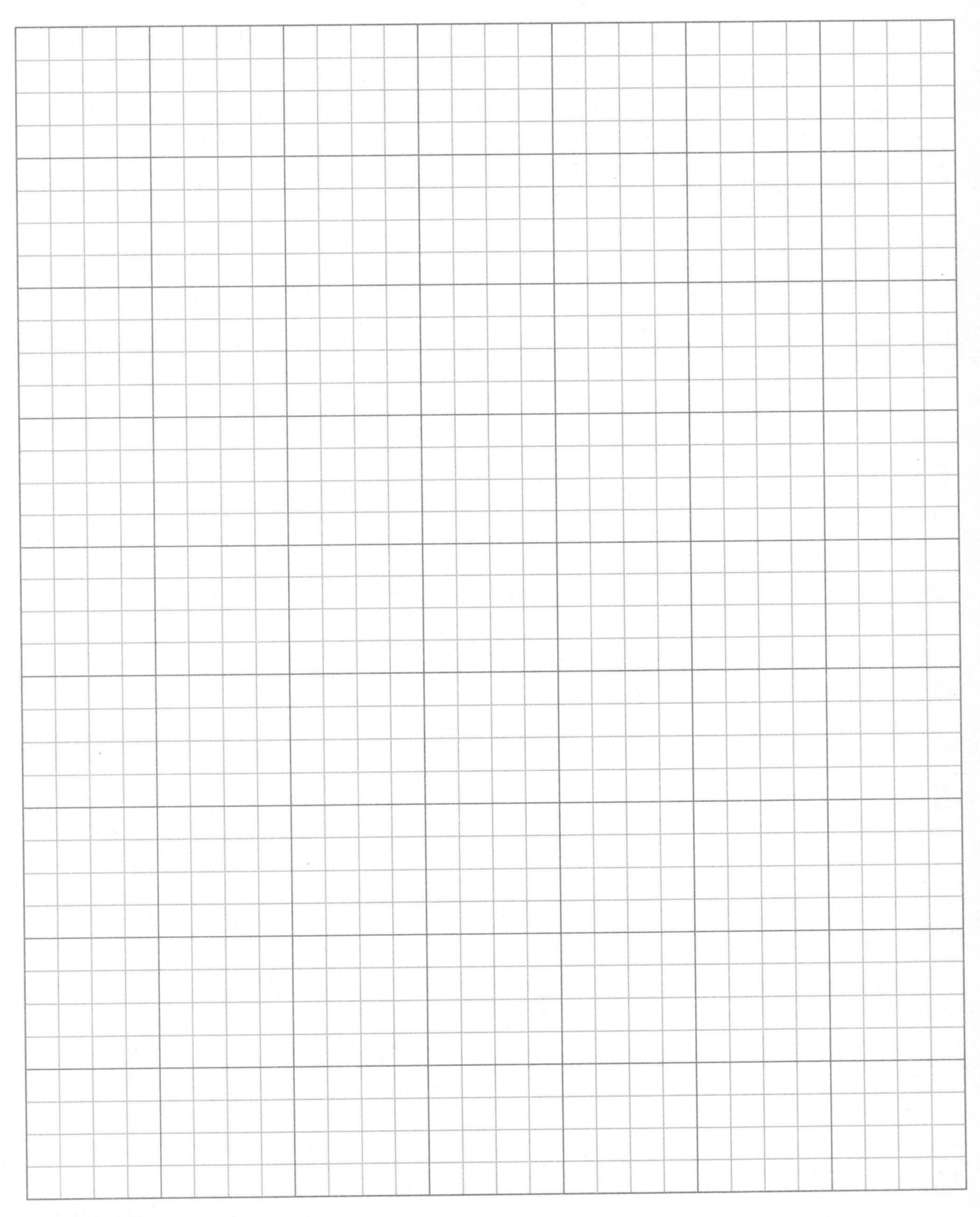

Activity 2-2 Worksheet *(Continued)*

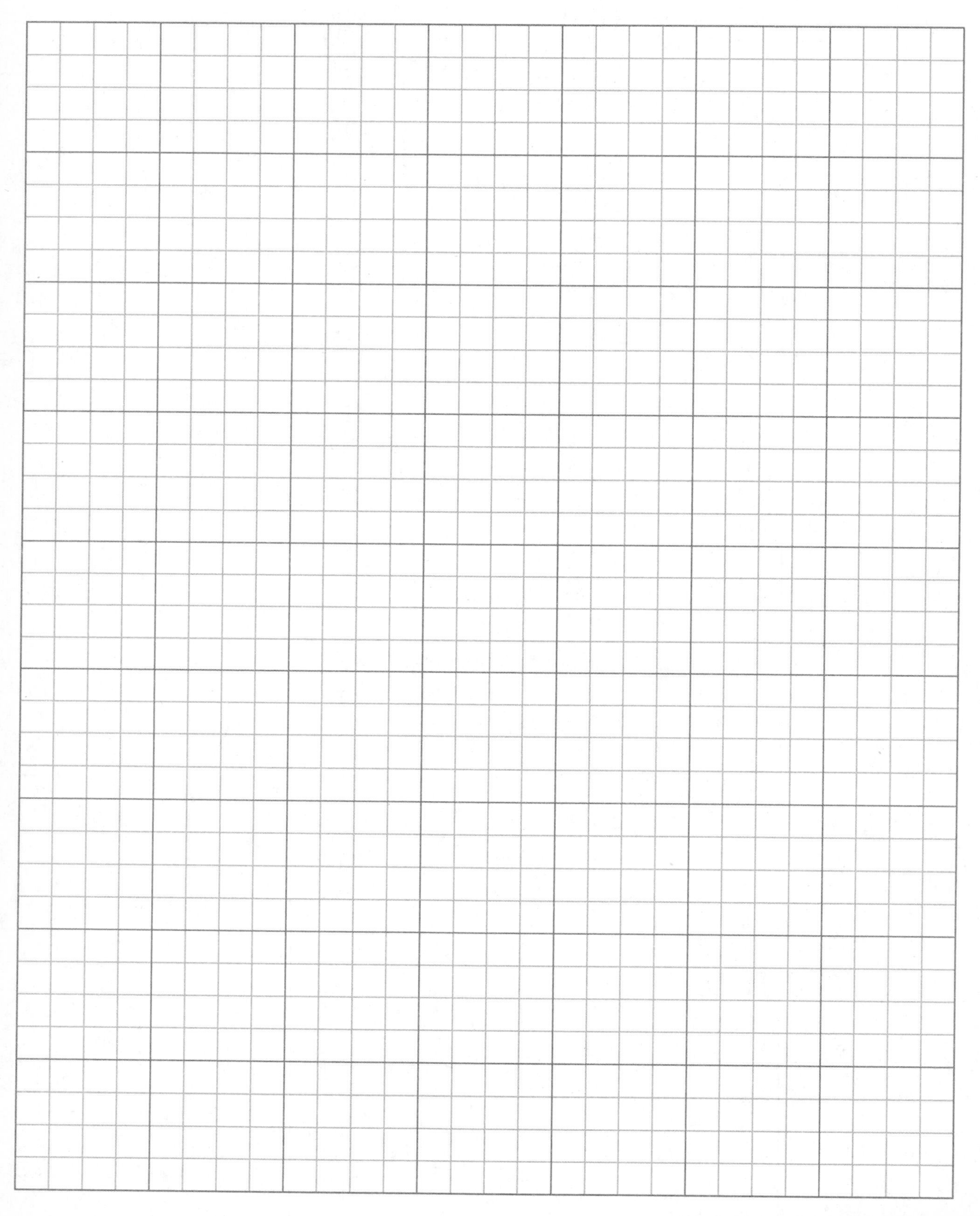

Name ______________________________ Date ____________ Class ___________

Activity 2-3

Idea Generation—Researching

When using research to generate ideas, engineers use historical and experimental research to identify possibly design solutions. In this activity, you will use research to generate ideas about a design problem.

Objective

After completing this activity, you will be able to:

- Use historical and/or experimental research to identify possible design solutions.

Materials

Pencil

Activity 2-3 Worksheet

Activity

In this activity, you will:

1. Use the Internet, your school library, or other sources to research the problems and solutions similar to yours in an effort to generate ideas on how to solve your problem.
2. Make sure you know all that you can about products like the one you are designing.
3. Use Activity 2-3 Worksheet to note ideas generated by research.

Reflective Questions

1. Did using research help generate ideas to solve your design problem?

____

____

2. What did you learn about similar problems that helped you generate ideas for your problem?

____

____

____

Activity 2-3 Worksheet

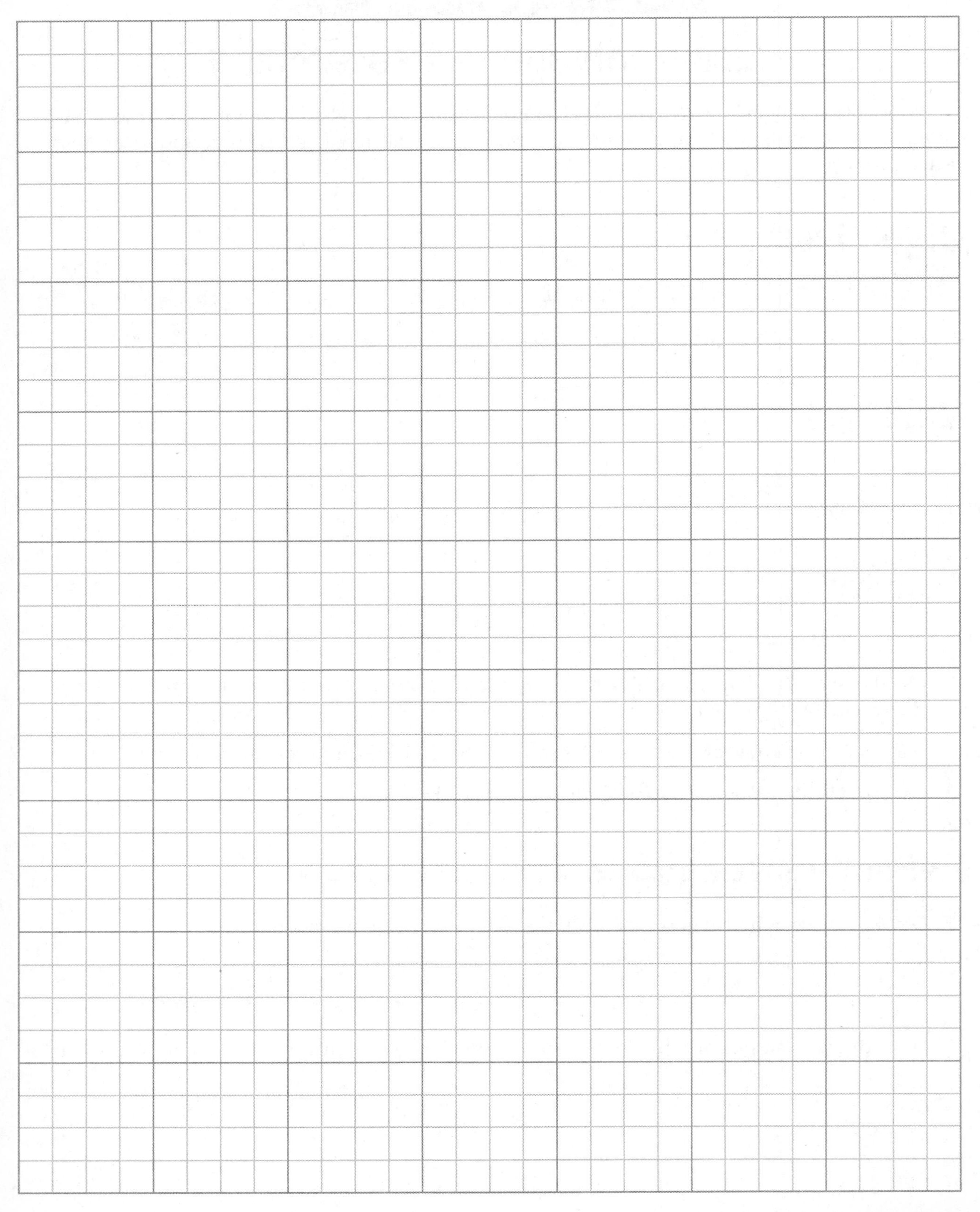

Name ______________________________ Date ______________ Class ____________

Activity 2-4

Solution Creation—Evaluating Solutions

In the solution creation step, engineers evaluate every idea from the idea generation step to decide on the idea that best solves the design problem. In this activity, you will evaluate each of the ideas from the previous steps and determine which idea best solves the problem.

Objective

After completing this activity, you will be able to:

- Evaluate solution ideas and determine which one bests solves the design problem.

Materials

Pencil

Activity 2-4 Worksheet

Activity

In this activity, you will:

1. Remove all ideas that obviously fail to solve the problem or fail to meet the criteria.
2. Using Activity 2-4 Worksheet, put ideas in either the *Ideas that could solve the problem* column or *Ideas that fail to solve the problem* column.
3. Evaluate each of the remaining ideas based on the problem and criteria.
4. Rank the remaining ideas from best to worst using the second part of Activity 2-4 Worksheet. Write some notes about each idea in the right column so you have a record of why you ranked the solutions in the order you did.

Reflective Questions

1. Did you keep most of the ideas as possible solutions? Why or why not?

   ____

   ____

2. Why did you rank the remaining ideas in the order you did? Is there one solution that seems better able to solve the problem?

   ____

   ____

Activity 2-4 Worksheet

Ideas that could solve the problem

Ideas that fail to solve the problem

Activity 2-4 Worksheet *(Continued)* Name ______________________________

Ideas that could solve the problem ranked from best to worst

Notes

Notes

Notes

Name ________________________________ Date ____________ Class ___________

Activity 2-5

Solution Creation—Communicating Solutions

After choosing the best possible idea to solve the problem, you must communicate that solution. In this activity, you will draw the idea in order to communicate your solution.

Objective

After completing this activity, you will be able to:

- Communicate the best solution to a design problem.

Materials

Computer with drafting or modeling software

Pencil

Activity 2-5 Worksheet

Activity

In this activity, you will:

1. Communicate the solution you chose.
2. Draw your solution in as much detail as possible so others can understand it and make it.
3. If you have access to a computer with drafting or modeling software, use that to communicate your solution.
4. If you do not have access to such software, use Activity 2-5 Worksheet to draw your solution.

Reflective Questions

1. How much detail did you feel you needed to include in your drawing in order to communicate the solution effectively?

   ____

   ____

2. Do you think any information in addition to your drawing would be needed in order to manufacture the solution? If so, what information is not included in your drawing?

   ____

   ____

Activity 2-5 Worksheet

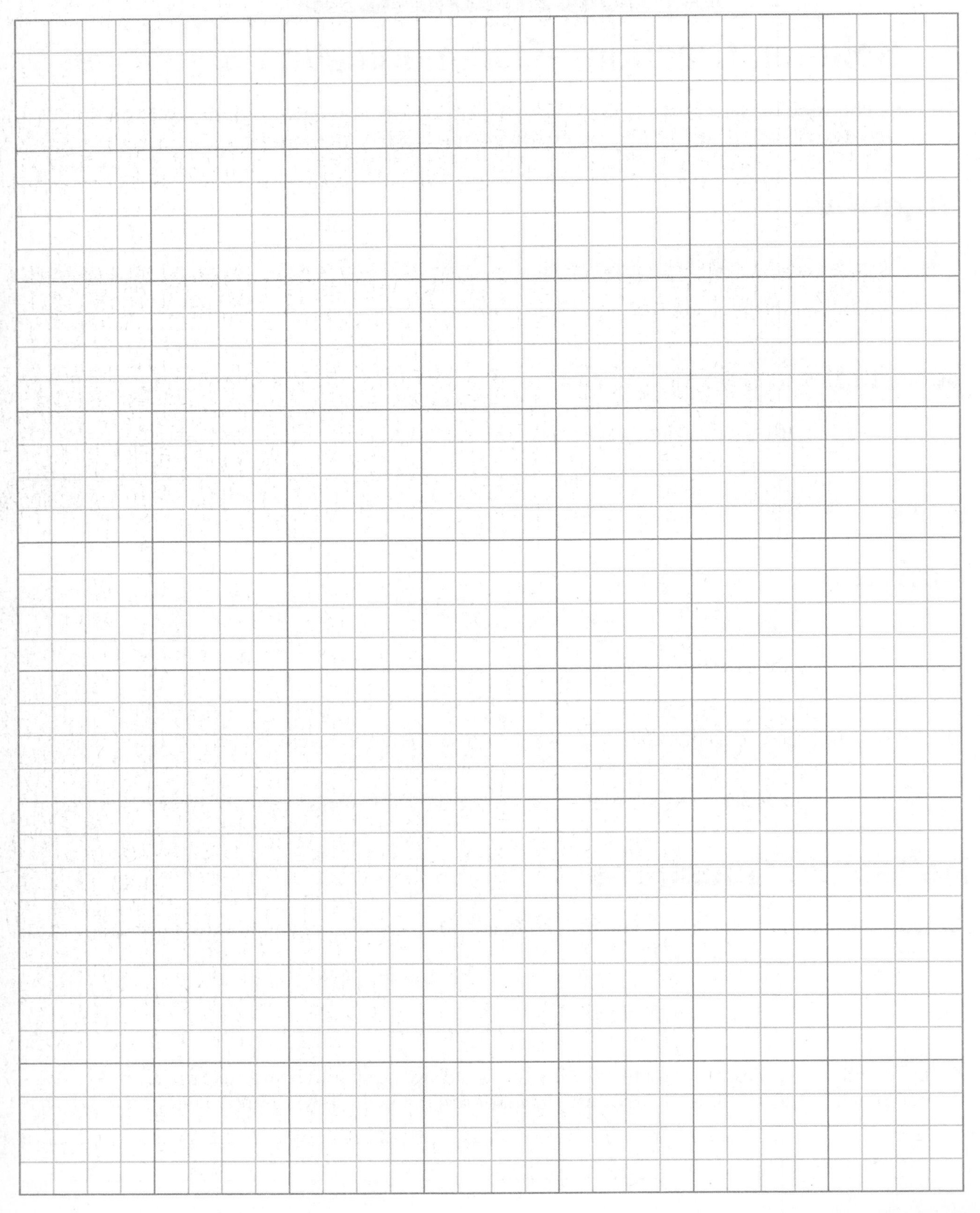

Name ______________________________ Date ____________ Class ___________

Activity 2-6

Test/Analysis

The purpose of the test/analysis step is to test and analyze the design solution against the identified criteria. Testing can be done using simulations, models, or prototypes. In this activity, you will test your design solution against the problem's criteria.

Objective

After completing this activity, you will be able to:

- Test and analyze a design solution against criteria.

Safety

If any tools and equipment are used to make a model or prototype, follow all safety procedures outlined by your instructor. Only use tools and equipment after you have been properly trained by your instructor and you are given permission.

Materials

Modeling materials

Modeling tools

Modeling equipment

Proper safety equipment

Activity

In this activity, you will:

1. Test and analyze your design solution against the problem statement and criteria. If you need to build a model or prototype of your solution, you will do so under the direct supervision of your instructor.
2. Conduct the necessary tests to evaluate your solution. For example, if you have designed a holder for CDs and DVDs, you will want to verify that the correct number of items fit in the holder and that it is adequately durable to perform its task. If you are designing a vehicle, it may need to be driven through a course or over obstacles. If cost is a constraint, you may need to develop a bill of materials and verify pricing.
3. If your design fails the tests and does not meet criteria, you can go back to a previous step, fix the problem, and test it again.

Reflective Questions

1. What kinds of tests did you use to evaluate your design solution against the criteria?

2. Did your design solution pass the tests and meet the criteria?

3. Does your design require any additional testing?

4. Can you think of any ways your design could be improved?

Name ______________________________ Date ____________ Class __________

Activity 2-7

Final Solution or Output

The purpose of the report design step is to communicate the design solution to the people who will make and purchase the product. Drawings and specifications sheets are used to communicate the size, shape, and materials.

Objectives

After completing this activity, you will be able to:

- Draw a product in enough detail that others can accurately build it.
- Write a specification sheet for a given product.

Materials

Pencil

Activity 2-7 Worksheet

Manual drawing tools

Computer with computer-aided drafting (CAD) software

Activity

In this activity, you will:

1. Using Activity 2-7 Worksheet, create a specification sheet to communicate your design solution.
2. Create dimensioned multiview drawings of your design solution using Activity 2-7 Worksheet.
3. Create an isometric drawing of your design solution using Activity 2-7 Worksheet.

Reflective Questions

1. Did you find your drawings from the solution creation step helpful in creating drawings of your design solution in this step? Why or why not?

2. Were your drawings in this step similar to your drawings from the solution creation step? What was different between the drawings from the two steps?

Name ________________________________

Activity 2-7 Worksheet

In the space provided below, list specifications for your product. This can include, but is not limited to, color, materials, finish, specific fasteners, adhesives, safety, and tolerance.

Activity 2-7 Worksheet *(Continued)*

In the space below, or using a CAD system, create dimensioned multiview drawings of your design solution.

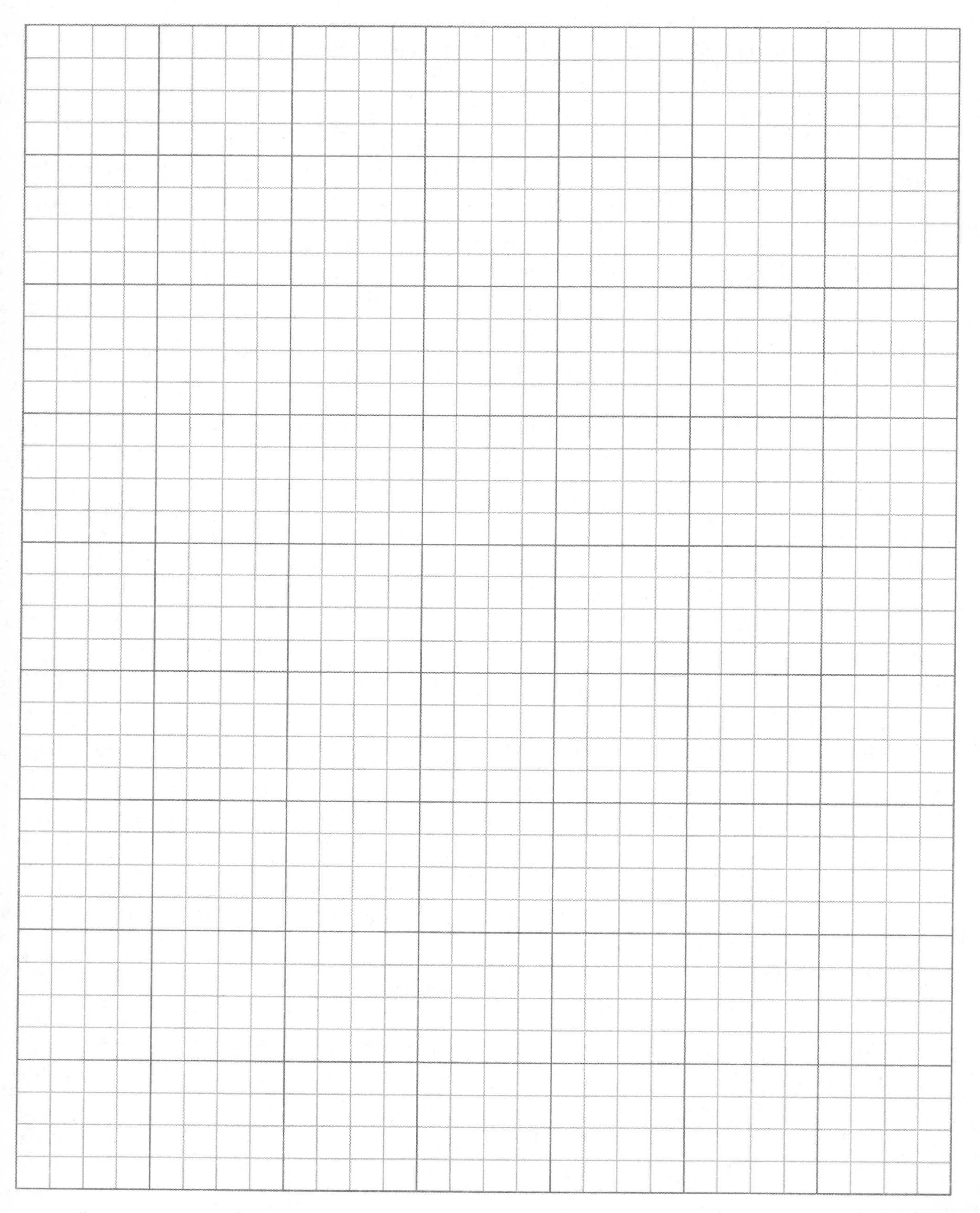

Activity 2-7 Worksheet *(Continued)* Name ______________________________

In the space below, or using a CAD system, create an isometric drawing of your design solution.

Additional isometric graph paper

Name ______________________________ Date ____________ Class ____________

Activity 2-8

Design Improvement

No design is perfect, and all designs can be improved in some way. In the design improvement step, engineers evaluate products that have passed testing and even gone into production to see if they can be made more effective, cheaper, safer, or more desirable. In this activity, you will look at your design solution from the previous steps and determine if it can be improved.

Objective

After completing this activity, you will be able to:

- Improve an existing design.

Materials

Pencil

Activity 2-8 Worksheet

Activity

In this activity, you will:

1. Think of any changes you could make to improve your design solution and write them on Activity 2-8 Worksheet.
2. Using Activity 2-8 Worksheet, draw a new solution based on possible changes for improvement.

Reflective Questions

1. When thinking of design changes, did you refer to your original ideas from the idea generation stage at all? Why or why not?

 ____

 ____

2. Did your design improvements help the solution meet more of the problem's criteria? If so, how?

 ____

 ____

Activity 2-8 Worksheet

Write down any changes you think could be made to improve your design.

Activity 2-8 Worksheet *(Continued)* Name ____________________

On the graph paper below, draw your solution to reflect the changes you are recommending to improve the design.

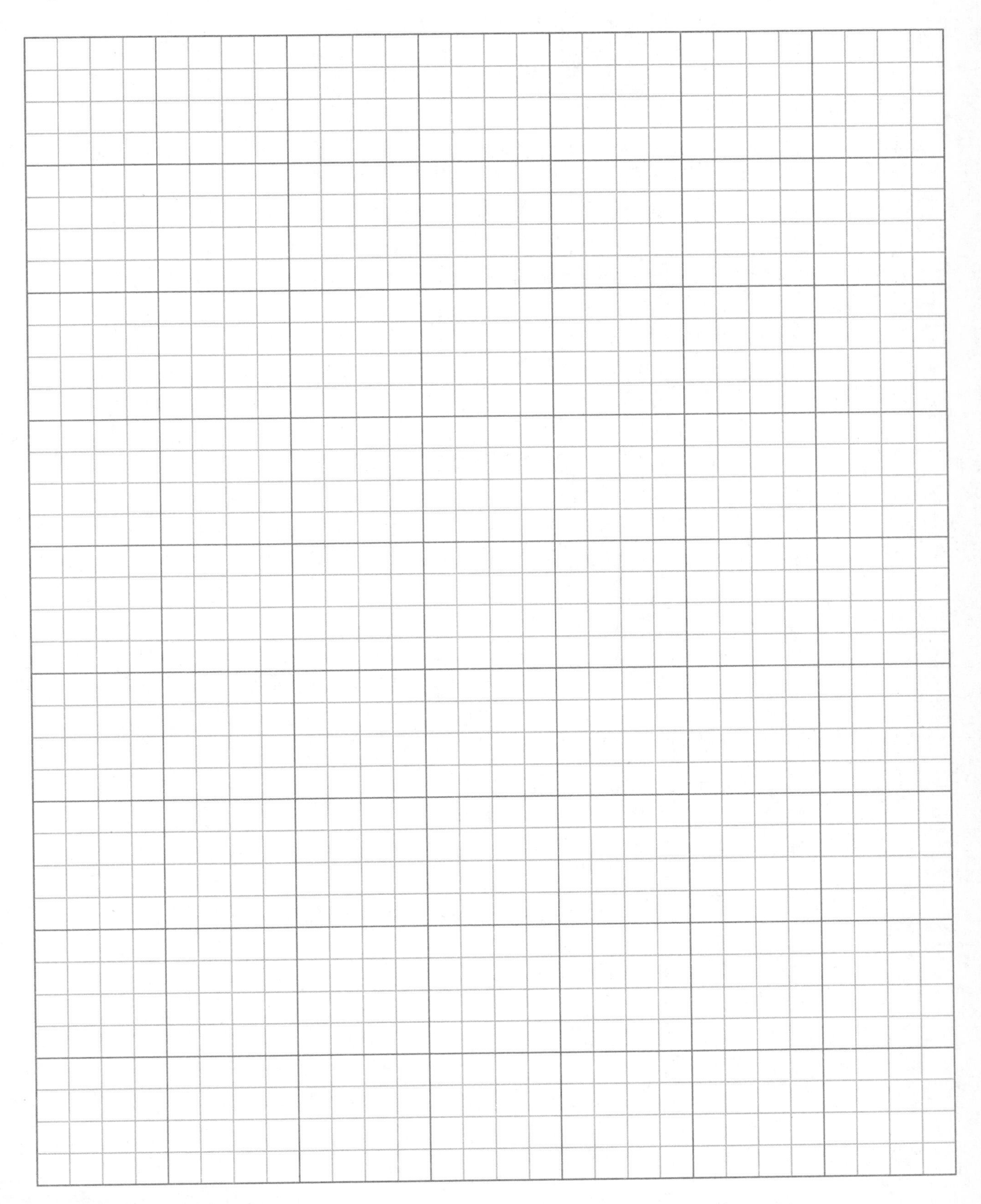

Additional graph paper

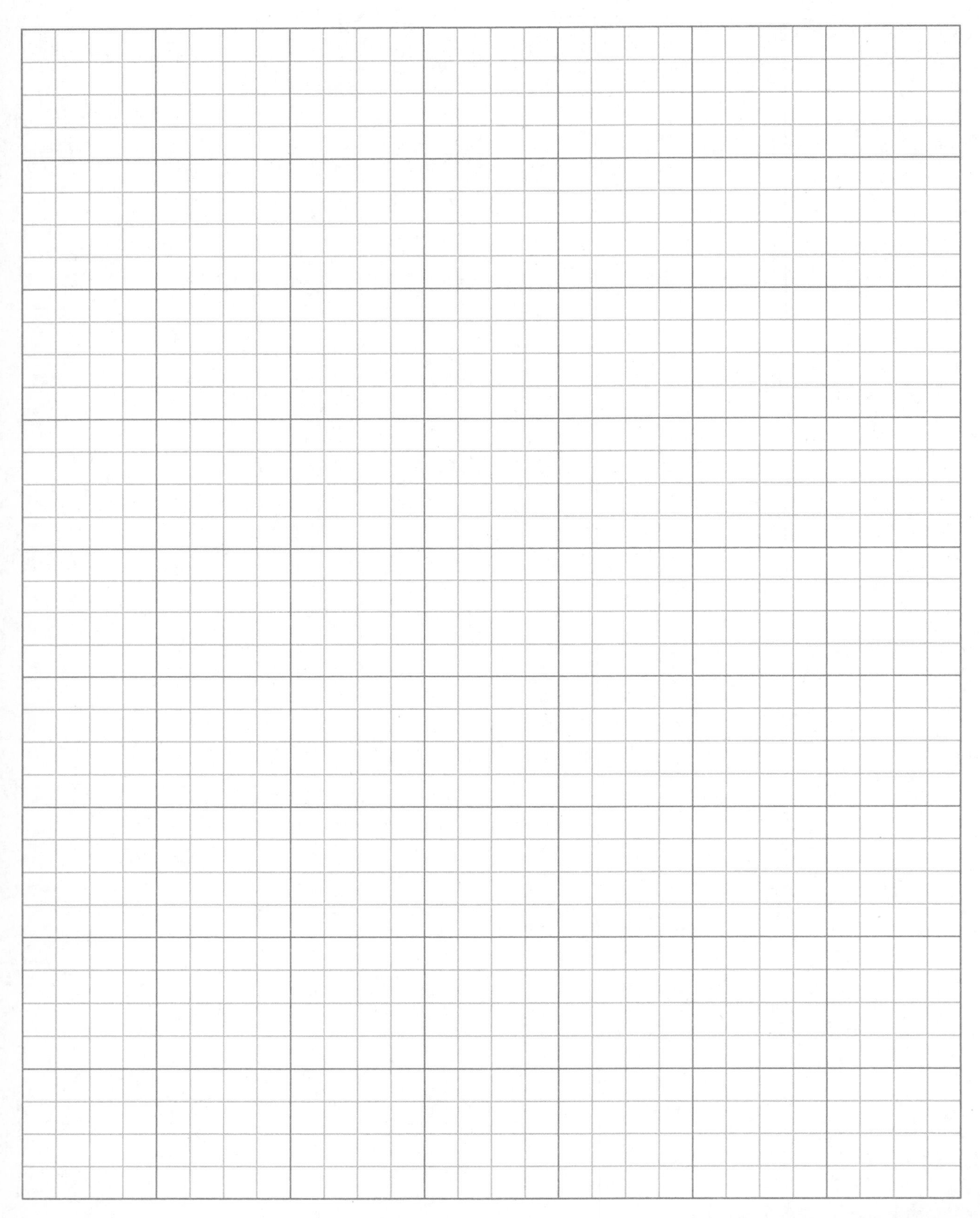

Chapter 3 Review

Defining Problems and Brainstorming

Name ______________________________ Date ____________ Class ____________

____________ 1. True or False? Engineers solve everyday problems.

2. Explain why engineers must clearly identify a problem.

____

____

______ 3. What is the last step in identifying problems?

A. Ask where the problem came from.

B. Determine what is and what is not the problem.

C. Look toward the desired solution.

D. State the problem in your own terms.

4. List the five factors that should be considered when developing a problem statement.

____

____

5. What are three things a problem statement should *not* do?

____

____

6. Explain the difference between criteria and constraints.

____

____

7. How are common constraints and specific constraints different from each other?

____

____

8. What four principles should be used to brainstorm a solution?

____

____

_______ 9. The _____ brainstorming technique is typically done individually.

A. free association

B. freewriting

C. future process

D. brainstorming web

_______________ 10. True or False? The ideas developed in a brainstorming session are considered final solutions.

Name ______________________ Date ____________ Class __________

Activity 3-1

Defining the Problem

Engineers solve problems. To correctly solve a problem, engineers must first be able to define the problem to solve. In this activity, you will define a problem to solve.

Objectives

After completing this activity, you will be able to:

- Define a problem.
- Determine what is and what is not part of a problem.

Materials

Paper

Activity 3-1 Worksheet

Activity

In this activity, you will:

1. Imagine you have noticed the car you are riding in is making a squeaking noise from the engine compartment.
2. Use Activity 3-1 Worksheet to define the problem with your vehicle.
3. Classify the potential problems at the top of the Worksheet in the correct spot.

Reflective Questions

1. How did you know if the vehicle component is or is not part of the problem?

    ____

    ____

2. What other methods can you use to define the problem?

    ____

    ____

    ____

Activity 3-1 Worksheet

Brake system	Engine
Wheels	Transmission
Battery	Rear seat bracket
Power window motor	Exhaust system

Is the problem

Is not the problem

Name ______________________________ Date ____________ Class ____________

Activity 3-2

Brainstorming

The brainstorming process is critical for all engineering disciplines. Engineers must be able to generate as many ideas as possible. Simple, elaborate, wild, and traditional ideas are all considered equally. In this activity, you will practice brainstorming.

Objectives

After completing this activity, you will be able to:

- Complete initial steps in the engineering design process.
- Conduct a brainstorming session.

Materials

Large surface to write, such as a whiteboard or chalkboard

Notebook

Writing utensils

Activity

In this activity, you will:

1. Imagine you have been hired by your local city to develop a new public transportation system.
2. In groups or individually, develop a transportation method to get you and your classmates to school by using the following process:
   A. Define the problem.
   B. List criteria and constraints.
   C. Brainstorm potential solutions using one of the methods described in the text.
   D. Document all your ideas.

5. Present your ideas to your teacher and the rest of the class.

Reflective Questions

1. What was the most difficult part of the brainstorming session?

____

____

____

2. Did you come up with different ideas you can combine together to solve the problem?

____

____

____

Chapter 4 Review

Researching Designs

Name ______________________ Date __________ Class __________

1. Who uses research and why?

____

____

2. What is a *thumbnail sketch*?

____

____

3. List the four basic steps of the sketching process.

____

____

____

4. Why do engineers do historical research?

____

____

____

5. List three forms of digital media used for research.

____

____

____

_______ 6. What is the main tool used in experimental research?

A. Books.

B. Tests.

C. Digital media.

D. Talking with other engineers.

7. What is feasibility?

____

____

____

________ 8. The _____ of a design is the level at which the design meets the intended outcome or effect.

A. feasibility

B. effectiveness

C. production

D. material properties

9. List four common considerations engineers use when creating a trade-off chart.

____

____

____

____

10. How do engineers share their final design selection with other stakeholders?

____

____

Name ______________________________ Date ____________ Class ___________

Activity 4-1

Sketching

Engineers use sketching as a way to help formulate ideas to solve problems. Most sketches start out as hand-drawn ideas. In this activity, you will practice sketching techniques.

Objective

After completing this activity, you will be able to:

- Use a four-step process to create sketches to a design problem.

Materials

Pencil

Drawing tools

Activity 4-1 Worksheet

Activity

In this activity, you will:

1. Imagine you are an engineer who is designing a portable sitting device. You need to design a chair that will be portable, will fit in the trunk of a car, will hold a 200-lb individual, and will be easily produced.
2. Use Activity 4-1 Worksheet to work through the sketching process. Use the four basic steps outlined in the text.

Reflective Questions

1. Could your designs be efficiently produced? Why or why not?

____

____

____

2. Who could you discuss these designs with to determine their potential effectiveness?

____

____

Activity 4-1 Worksheet

Visualize the solution

Outline the solution

Block out the solution

Detail the solution

Name ________________________________ Date ____________ Class __________

Activity 4-2

Problem Solving

Engineers solve problems. In this activity, you will work through multiple steps of the design process by researching and analyzing potential designs.

Objectives

After completing this activity, you will be able to:

- Draw sketches.
- Research solutions.
- Complete a trade-off matrix.

Materials

Paper

Drawing tools

Activity 4-2 Worksheet

Activity

In this activity, you will:

1. Imagine you are an engineer who must develop a new method to automatically sort books at your school library.
2. Work through the problem by developing sketches, researching potential solutions, and completing the trade-off matrix.
3. Use Activity 4-2 Worksheet to complete the following steps:
   A. Develop sketches.
   B. Research potential solutions.
   C. Complete the trade-off matrix.
4. Present your ideas to the rest of your class.

Reflective Questions

1. What types of research methods did you use? Why?

____

____

____

____

2. How can research help improve your design?

____

____

____

Activity 4-2 Worksheet

Sketches

Activity 4-2 Worksheet *(Continued)*

Research Notes

Activity 4-2 Worksheet *(Continued)*　　Name ______________________________

Research Notes							
Solution	Appearance	Manufacturability	Feasibilty	Low maintenance	Marketability	Environmentally friendly	Total

Notes

Name ______________________________ Date ____________ Class __________

Activity 4-3

Spinoffs

Engineers often unintentionally solve problems. Sometimes, products designed for a specific purpose work in another situation. The National Aeronautics and Space Administration (NASA) has engineered many solutions that have become useful in our everyday lives. These solutions are called *spinoffs*. In this activity, you will research NASA spinoffs.

Objectives

After completing this activity, you will be able to:

- Research NASA spinoff technology.
- Describe a NASA spinoff to your teacher and classmates.

Materials

Computer with Internet access

Activity

In this activity, you will:

1. Use an Internet search engine to find NASA's spinoff website.
2. Review three different spinoffs.
3. Select one spinoff to share with your classmates.
4. Research the following information about the spinoff:
   A. When was it invented?
   B. Who invented the spinoff?
   C. What was the tool initially designed for?
   D. How is the spinoff used today?
5. Present your findings to the rest of your class.

Reflective Questions

1. What was the most surprising spinoff? Why did you find it surprising?

2. Look at the tools in your classroom. Are there possible other uses for them?

Chapter 5 Review

Communicating Solutions

Name ______________________________ Date ______________ Class ____________

______________ 1. True or False? Engineering drawings are the initial sketches of a design solution.

2. Describe how engineers use the following drawings:

   A. Detail Drawings.

   ____

   ____

   B. Assembly Drawings.

   ____

   ____

   ____

   C. Schematic Drawings.

   ____

   ____

3. What do orthographic drawings describe?

____

____

4. Why do engineers need to be able to visualize a solution?

____

____

____

5. What type of orthographic drawing is used to show flat pieces?

____

6. In what situation do engineers use a three-view drawing?

____

____

Matching

_______ 7. Type of drawing that uses angles equal to each other and are always drawn with standard angles.

_______ 8. Type of drawing that shows an object for a specific point of view

_______ 9. Type of drawing that highlights one side of an object.

A. Isometric drawing

B. Oblique drawing

C. Perspective drawing

10. What is the most unique part of a perspective drawing?

____

11. Why are symbols used in engineering drawings?

____

____

12. What are the five primary line types used in an engineering drawing?

____

13. What does it mean to have a drawing that is drawn at full scale?

____

14. What are two types of lines used by engineers when dimensioning an object?

____

____

15. Why might a company have its own procedures and guidelines for drawings?

____

____

Name ______________________________ Date ______________ Class ____________

Activity 5-1

Dimensioning

Engineers dimension all their engineering drawings. You use different types of dimensions for the many different shapes which are dimensioned. Engineers are guided by specific guidelines used in their industry or specific company. In this activity, you will dimension the provided shapes.

Objective

After completing this activity, you will be able to:

- Dimension different shapes.

Materials

Engineer's scale
Ruler
Pencil
Activity 5-1 Worksheet

Activity

In this activity, you will:

1. Provide dimensions for the objects in Activity 5-1 Worksheet.
2. You should follow the guidelines suggested in the textbook and the guidelines provided by your instructor.

Reflective Questions

1. Are there specific dimensions that are critical for the design?

____

____

2. Should you repeat dimensions in a multiview drawing? Why or why not?

____

____

3. Which drawing do you think most clearly shows the object? Why?

____

____

Activity 5-1 Worksheet

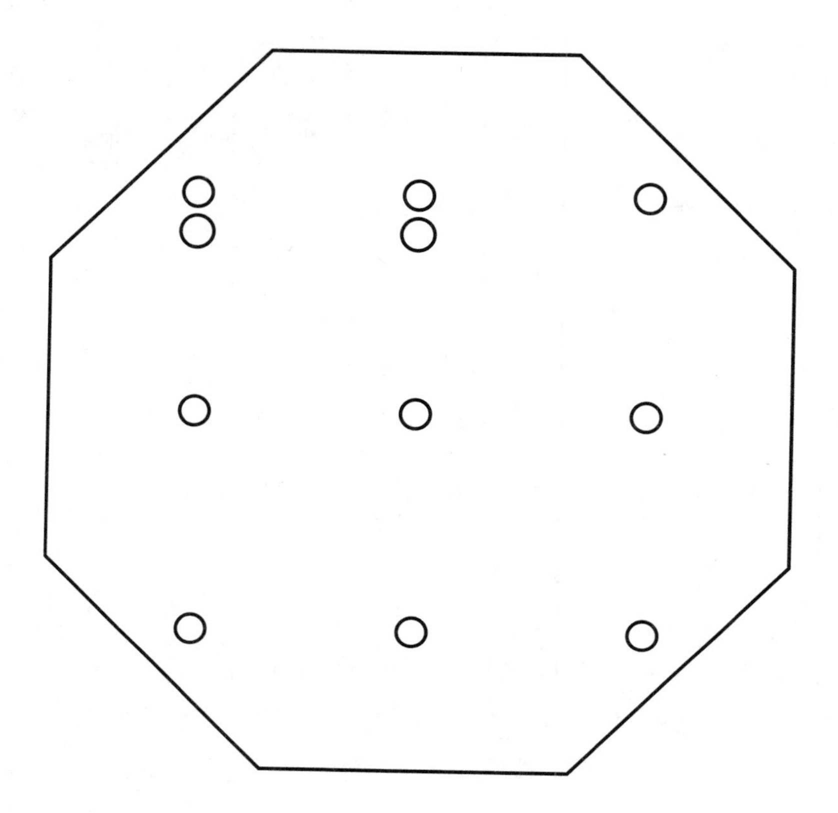

1/4″ THK PLATE

One-View Drawing

Activity 5-1 Worksheet *(Continued)* Name ____________________

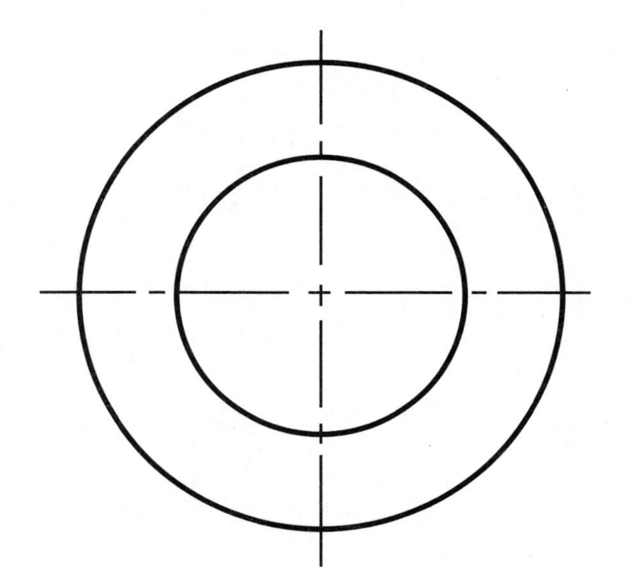

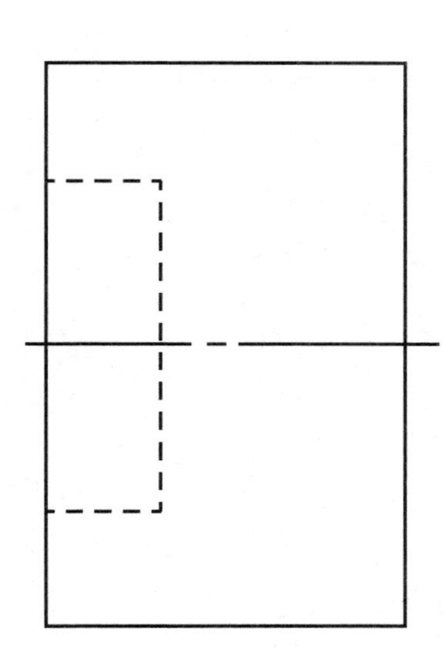

Two-View Drawing

Activity 5-1 Worksheet *(Continued)*

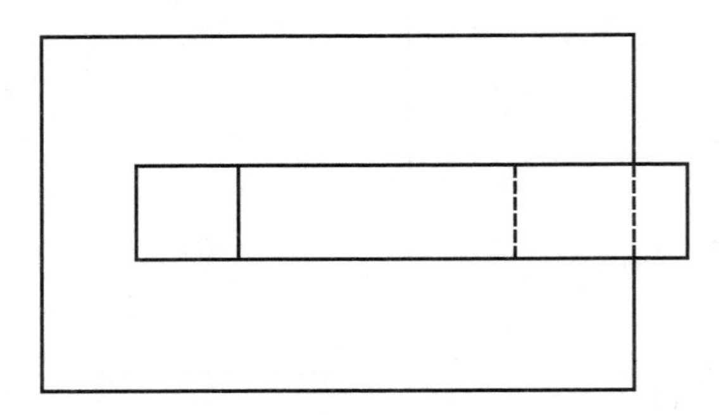

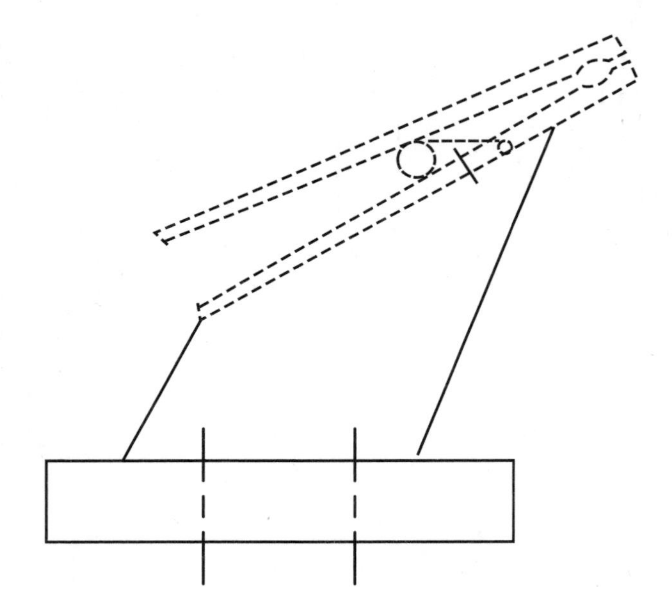

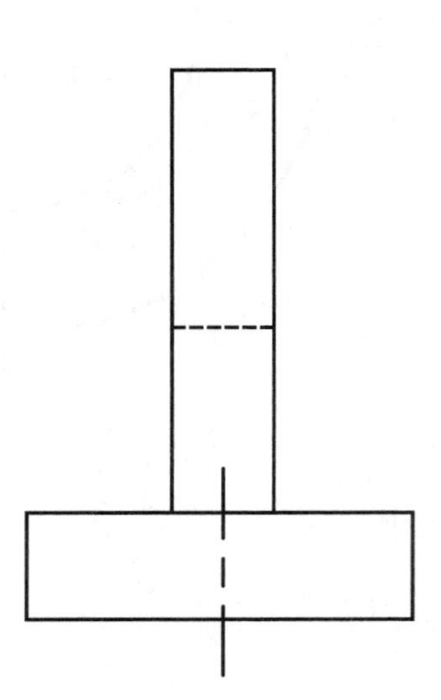

Three-View Drawing

Name ______________________________ Date ____________ Class ____________

Activity 5-2

Orthographic Drawings

Engineers use different types of orthographic drawing techniques to produce drawings that show details of their designs. Orthographic drawings can have one, two, or three views, depending on the shape of the object. In this activity, you will create orthographic drawings using different techniques.

Objectives

After completing this activity, you will be able to:

- Draw the missing view in an orthographic drawing.
- Dimension an orthographic drawing

Materials

Drawing tools

Pencil

Activity 5-2 Worksheet

Activity

In this activity, you will:

1. Identify the missing view(s) in each drawing.
2. Use the information from the provided view to create the missing view(s).
3. Include different line types as appropriate.
4. Provide appropriate dimensioning to the drawings where necessary.

Reflective Questions

1. Why were the missing views needed?

2. Did you need to add dimensions to the object? If so, why?

Name ________________________________

Activity 5-2 Worksheet

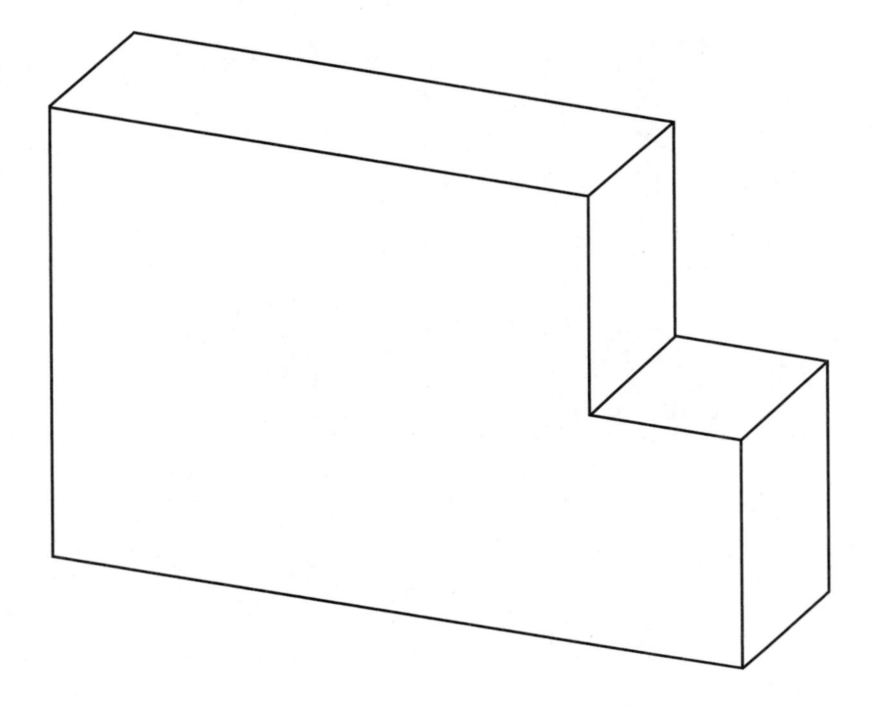

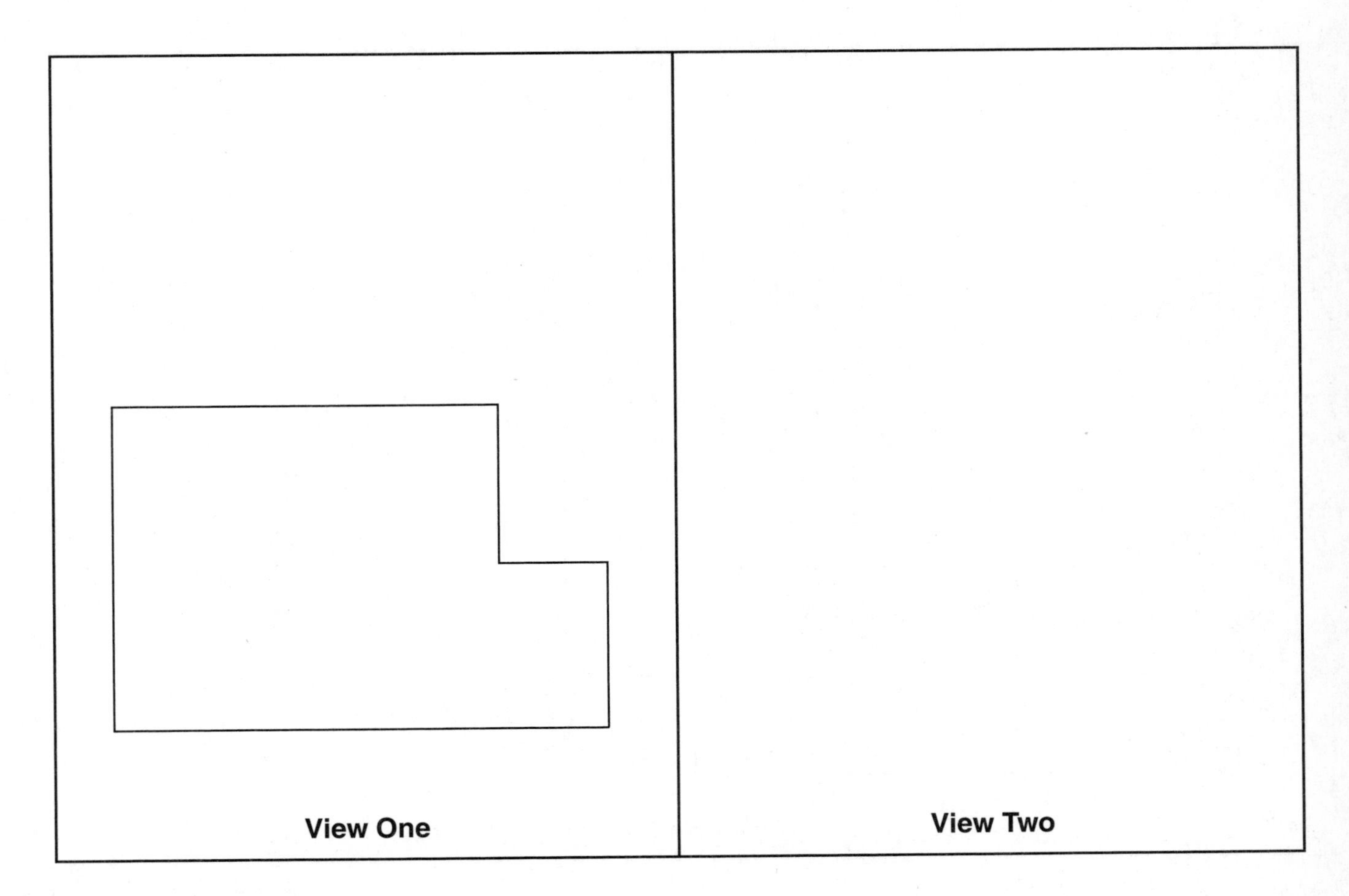

Activity 5-2 Worksheet *(Continued)*

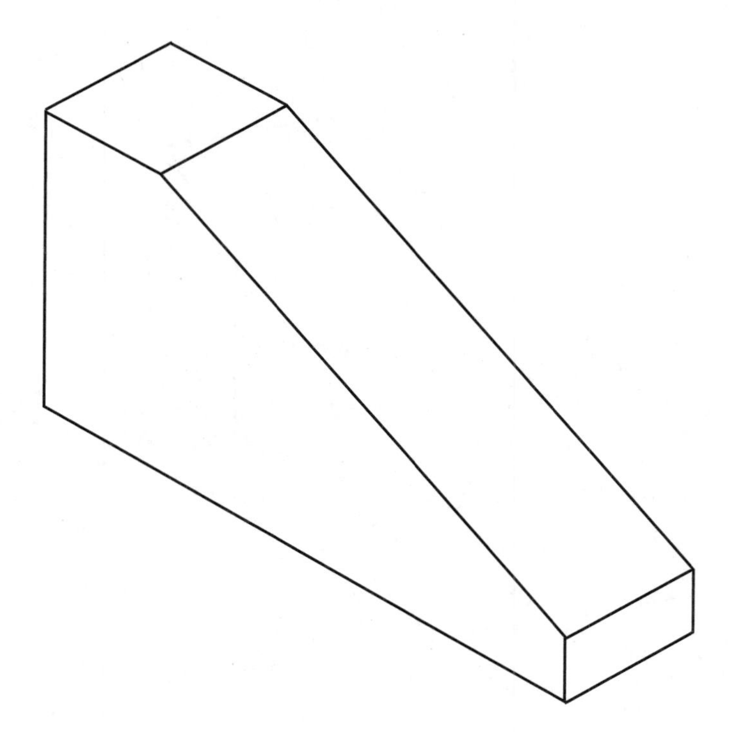

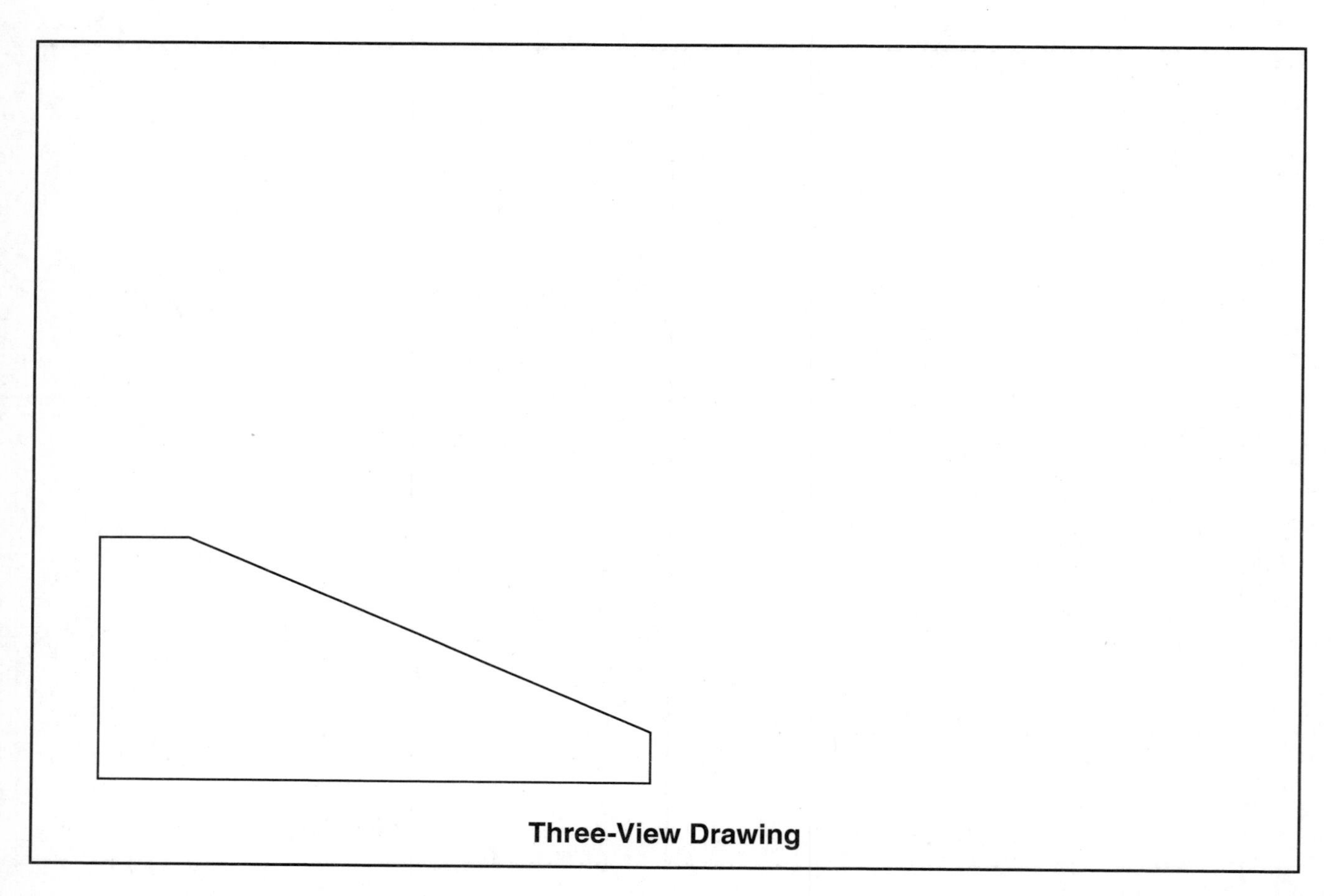

Three-View Drawing

Name ______________________________________ Date ____________ Class __________

Activity 5-3

Perspective Drawings

Engineers use perspective drawings to show an object from a specific point of view. Engineers may use a one-, two-, or three-point perspective. In this activity, you will draw a perspective drawing of an object provided by your teacher.

Objectives

After completing this activity, you will be able to:

- Determine the appropriate type of perspective drawing to communicate a design.
- Draw a perspective drawing.

Materials

Drawing tools

Activity 5-3 Worksheet

Activity

In this activity, you will:

1. View the object provided by your teacher.
2. Visualize the perspective drawing.
3. Select the appropriate method (one-, two-, or three-point perspective drawing).
4. Using Activity 5-3 Worksheet, complete the perspective drawing using the method suggested in the textbook.
5. Communicate the design with your teacher.

Reflective Questions

1. What details are you able to show with a perspective drawing?

2. Why would an engineer want to use a perspective drawing?

3. How did you select your vanishing point?

Name ___________________________________

Activity 5-3 Worksheet

Perspective Drawing

Additional sketching paper

Chapter 6 Review

Modeling, Testing, and Final Outputs

Name ______________________________ Date ____________ Class ___________

1. Why is mathematical modeling used?

____

____

Matching

_____ 2. Mass.

_____ 3. Length.

_____ 4. Energy.

_____ 5. Torque.

_____ 6. Area.

_____ 7. Volume.

_____ 8. Angle.

A. Foot-pound
B. Square inches
C. Degrees
D. Yards
E. Grams
F. Cubic inches
G. Pound-foot

9. What is the difference between a mock-up and a prototype?

____

____

____

10. Identify four different engineering fields and describe the type of computer modeling software they use.

____

____

____

____

_____________________ 11. True or False? Assumptions make predictive analysis more effective.

12. List five different questions engineers might ask about the function of a design.

____

____

____

____

____

13. List five different questions engineers might ask about the aesthetics of a design.

____

____

____

____

____

14. What are three types of final outputs for an engineering project?

____

____

____

_______ 15. _____ is a method of determining the properties or function of a device by taking it apart and looking at its operation structure.

A. Predictive analysis

B. Engineering economics

C. Geospatial modeling

D. Reverse engineering

Name ____________________________ Date ____________ Class ___________

Activity 6-1

Product Evaluation

Engineers evaluate products to determine their effectiveness. All the products you use on a daily basis have been evaluated using different criteria. In this activity, think of a product you use every day, and evaluate the product on each of the five criteria used by engineers.

Objectives

After completing this activity, you will be able to:

- Evaluate engineering design.
- Communicate evaluation of engineered products.

Materials

Various engineered products used daily by students

Pencil

Activity 6-1 Worksheet

Activity

In this activity, you will:

1. Select a product to analyze.
2. Use the guidelines and questions in the textbook to analyze the product based on the five criteria: function, fit, aesthetics, safety, and environmental impact. Use Activity 6-1 Worksheet for taking notes.
3. Develop questions not suggested in the text.
4. Provide a 1–2 page written summary on the evaluation of your product.

Reflective Questions

1. Would you make changes to the design? If so, what changes?

2. What are some questions not given in the textbook that you used to analyze the product design?

Name ______________________________

Activity 6-1 Worksheet

Function

____

____

____

____

____

Fit

____

____

____

____

____

Aesthetics

____

____

____

____

____

Safety

____

____

____

____

____

Environmental Impact

____

____

____

____

____

Notes

Name ______________________________ Date ____________ Class ____________

Activity 6-2

Reverse Engineering

Reverse engineering is used to learn more about or to make changes to an existing design. When a design is complete, or if there are problems with a design, engineers reverse engineer the design by taking it apart and looking at each piece to make improvements. In this activity, you will choose a device to work through the reverse engineering process.

Objective

After completing this activity, you will be able to:

- Use reverse engineering to suggest improvements in a tool design.

Materials

Product to analyze

Tools for disassembly

Activity

In this activity, you will:

1. Select a device to reverse engineer. You may want to select a broken device that you can easily take apart without damaging.
2. Disassemble the device to understand how it works.
3. Investigate how the product works, and make notes in your engineering notebook about the function, fit, aesthetics, safety, and environmental impact of the design.
4. Consider ways this device could either be repaired or improved by changing the design.
5. Work through the engineering design process to make changes to the product. You may either redesign the product on paper or make the actual improvements or repairs on the product. Work with your teacher to determine the approach for your project.

Reflective Questions

1. What could you improve on the design? Why?

2. Could your changes be used to modify the production of the product? How?

Name ________________________________ Date ____________ Class ___________

Activity 6-3

Final Project Report

Engineers communicate their development of an engineering solution through a final project report. The final project report provides information about each step of the process. In this activity, you will create a final project report for an engineered product given to you by your teacher.

Objectives

After completing this activity, you will be able to:

- Develop an final project report.
- Communicate engineering design information.

Materials

Engineered project

Measuring tools

Sketching paper

Pencil

Activity

In this activity, you will:

1. View the object provided by your teacher.
2. Visualize the different steps taken in the development of the product.
3. Based on the project summary information in the text, create an outline similar to what engineers would use to communicate their engineered design.
4. Create a presentation for your class based on the completed final project report.

Reflective Questions

1. What factors did the engineer need to consider in designing the product?

____

____

____

____

2. What design components should be reconsidered through a reverse engineering process? Why?

____

____

____

____

Chapter 7 Review

Materials Engineering

Name ______________________ Date __________ Class __________

1. List two types of projects that materials engineers would design.

____

____

2. List two professional organizations for materials engineers.

____

____

______________ 3. Metals have a(n) _____ structure.

Matching

_______	4. Metal	A. Plywood
_______	5. Ceramic	B. Clay
_______	6. Polymer	C. Brass
_______	7. Composite	D. PVC

______________ 8. A mixture of two metals is known as a(n) _____.

______________ 9. _____ is a ceramic that does not have a crystalline structure.

10. List the two main types of plastics.

____

____

______________ 11. Wood is a natural _____.

______________ 12. A(n) _____ is a material that interacts with living systems.

_______ 13. Stress, strain, and strength are _____ properties

A. mechanical

B. physical

C. chemical

D. electrical

_______ 14. Flammability is a(n) _____ property.

A. mechanical

B. physical

C. chemical

D. electrical

_______ 15. Conductivity and resistivity are _____ properties

A. mechanical

B. physical

C. chemical

D. electrical

16. Describe the difference between destructive and nondestructive material tests.

____

____

____

_______ 17. An example of a destructive test is a(n) _____.

A. visual examination

B. radiography test

C. tensile test

D. ultrasonic test

_______________ 18. Nanotechnology is the design of products at the scale of one _____ of a meter.

_______________ 19. Researchers are experimenting with using _____ to carry medicine to specific cells in the human body.

_______________ 20. True or False? Manufacturability is not a consideration when selecting materials.

Name ______________________ Date __________ Class __________

Activity 7-1

Material Types

Understanding the types and uses of materials is fundamental for a materials engineer. Throughout your daily life, you interact with a number of materials. In this activity, you will identify several of the materials with which you come in contact.

Objective

After completing this activity, you will be able to:

- Provide examples of various materials

Materials

Pencil

Activity 7-1 Worksheet

Activity

In this activity, you will:

1. Be aware of the materials and products that you touch throughout the day.
2. Use Activity 7-1 Worksheet to list products within each material that you have interacted with.
3. List the characteristics of the materials on Activity 7-1 Worksheet.
4. Present your findings to the rest of the class.

Reflective Questions

1. Were you surprised at the amount of materials that you use on a daily basis?

____

____

2. What did you learn about materials types from this assignment?

____

____

____

Activity 7-1 Worksheet

Material	Product	Characteristic
Example: Polymer	Toothbrush	Solid, but a bit flexible, colored white
Metal		
Metal		
Metal		
Metal		
Metal		
Ceramic		
Ceramic		
Ceramic		
Ceramic		
Ceramic		
Polymer		
Polymer		
Polymer		
Polymer		
Polymer		
Composite		
Composite		
Composite		
Composite		
Composite		
Other		
Other		

Name ______________________________ Date ____________ Class ___________

Activity 7-2

Material Types Follow-Up

In the previous activity, you determined the materials in a number of products that you interact with daily. In this activity, you will consider other material choices for those products.

Objective

After completing this activity, you will be able to:

- Determine appropriate material choices.

Materials

Pencil

Completed Activity 7-1 Worksheet

Activity 7-2 Worksheet

Activity

In this activity, you will:

1. Review your completed Activity 7-1 Worksheet.
2. Select four of the products you wrote down on Activity 7-1 Worksheet, and fill those in on Activity 7-2 Worksheet.
3. For each product, select two alternative materials.
4. Describe the advantages and disadvantages if the product were to be made from the alternative materials.
5. Circle the material you feel is the best material for the product.

Reflective Questions

1. How did you determine the best material for a product? What criteria did you think about?

2. How do you believe that material engineers determine the best materials for a product?

Name ____________________

Activity 7-2 Worksheet

Product #1:	Original Material:	Characteristics:
Alternative Material #1:	Advantages:	Disadvantages:
Alternative Material #2:	Advantages:	Disadvantages:

Product #1:	Original Material:	Characteristics:
Alternative Material #1:	Advantages:	Disadvantages:
Alternative Material #2:	Advantages:	Disadvantages:

Product #1:	Original Material:	Characteristics:
Alternative Material #1:	Advantages:	Disadvantages:
Alternative Material #2:	Advantages:	Disadvantages:

Product #1:	Original Material:	Characteristics:
Alternative Material #1:	Advantages:	Disadvantages:
Alternative Material #2:	Advantages:	Disadvantages:

Notes

Name ________________________________ Date ______________ Class ____________

Activity 7-3

Material Properties

In this activity, you will research properties and common uses of a specific material.

Objective

After completing this activity, you will be able to:

- Describe various material properties.

Materials

Computer or library access for research purposes

Presentation software

Activity

In this activity, you will:

1. Select a specific material.
2. Research the material to determine at least one material property in each the following categories:
   A. Physical properties.
   B. Mechanical properties.
   C. Electrical and magnetic properties.
   D. Chemical properties.
   E. Thermal properties.
   F. Optical and acoustical properties.
3. Find a number of common uses for the material.
4. Create a presentation of your research findings using presentation software.
5. Present your material and its properties to the rest of the class.

Reflective Questions

1. What was the most interesting piece of information that you learned about the material?

2. What types of uses would your material *not* be suitable for?

Name ________________________________ Date ____________ Class __________

Activity 7-4

Material Testing Device

Materials engineers often need to understand the properties of a specific material to understand how they would work in a product. To do this, they may have to test a material. In this activity, you will design and build a device to test a material property.

Objective

After completing this activity, you will be able to:

- Determine a property of a material through the design and use of a testing device.

Safety

Safety requirements will depend on the nature of the materials used in this activity and will be outlined by your instructor.

Materials

Materials will vary based on the nature of the testing device being built

Activity

In this activity, you will:

1. Determine a material property (such as thermal resistance, light reflection, resistivity, or tensile strength) for which you would like to build a testing device.
2. Conduct research to better understand the selected material property and to determine common tests that are associated with the material property.
3. Design a material testing device. There are several items you must consider as you are designing your device:
   A. There must be a method of ensuring the tests are consistent each time they are performed.
   B. The device should be able to test a range of materials.
   C. There should be a method of measuring the results of the test.
4. Build your testing device.
5. Test the testing device using several different materials.
6. Make any needed changes.

7. Create a log sheet that could be used with your testing device to record the necessary data.
8. Demonstrate your testing device to the rest of the class.

Reflective Questions

1. What did you learn about material testing from designing and building a material testing device?

____
____
____
____
____
____

2. What changes could be made to make your testing device more accurate?

____
____
____
____
____
____

Name ______________________________ Date ______________ Class ____________

Activity 7-5

Materials Engineering Design

In this activity, you will solve a materials engineering design problem using the engineering design process.

Objectives

After completing this activity, you will be able to:

- Identify a materials engineering problem.
- Complete a materials engineering design problem.

Safety

Safety requirements will depend on the nature of the materials used in this activity and will be outlined by your instructor.

Materials

Materials will vary based on the design problem being solved

Engineering notebook

Pencil

Activity

In this activity, you will:

1. Determine a problem a materials engineer would be asked to solve.
2. Utilize the steps of the engineering design process presented in Chapter 2 to solve the problem:
   A. Problem definition
   B. Idea generation
   C. Solution creation
   D. Testing/analyzing
   E. Final solution or output
   F. Design improvement
3. Throughout the process, keep an engineering notebook to record your notes, drawings, and findings.
4. Present your problem and solution to the rest of the class.

Reflective Questions

1. What did you learn about materials engineering from conducting the design process?

2. What other materials engineering problems could be solved using an engineering design process?

Chapter 8 Review

Electrical Engineering

Name ______________________ Date __________ Class __________

1. List two types of projects electrical engineers might design.

____

____

2. What is the requirement for an entry-level electrical engineer?

____

3. What is the function of IEEE?

____

____

Matching

_______ 4. Electron.

_______ 5. Proton.

_______ 6. Neutron.

A. Neutral
B. Positive
C. Negative

_______________ 7. In the electron flow theory, electricity flows from _____ to _____.

Matching

_______ 8. Electricity from chemicals.

_______ 9. Electricity from magnets.

_______ 10. Electricity from sunlight.

A. Solar
B. Generator
C. Battery

_______ 11. _____ is expressed as electromotive force (EMF).

A. Current

B. Resistance

C. Voltage

D. Power

____________________ 12. The measure of the amount of electricity equal to 6.24×10^{18} electrons are _____.

_______ 13. _____ is a measurement of the number of electrons passing a given point in a given amount of time.

A. Current

B. Resistance

C. Voltage

D. Power

____________________ 14. The letters _____ and _____ represent voltage and current respectively.

_______ 15. _____ is the amount of opposition to electron flow.

A. Current

B. Resistance

C. Voltage

D. Power

16. Explain Ohm's law, and list the three related formulas in the space below.

____

____

____

____

____________________ 17. Solve for the total resistance in a series circuit consisting of a 50 Ω, 100 Ω, and 200 Ω resistor.

_______ 18. _____ circuits have more than one path for current flow.

A. Series

B. Parallel

C. Short

D. Energized

Chapter 8 Review *(Continued)* Name ______________________________

________ 19. In _____ circuits, voltage is equal across all loads.

A. series

B. parallel

C. series-parallel

D. energized

________________ 20. What is the total resistance in a circuit where three 25 Ω resistors are wired in parallel?

21. List three commonly used conductors.

____

____

____

22. List three commonly used insulators.

____

____

____

________________ 23. Insulators have very _____ resistance and conductors have very _____ resistance.

Matching

________ 24. Create electrical signals based on conditions.

________ 25. Function like fluorescents, but fit standard light sockets.

________ 26. Entire circuits enclosed in a plastic case.

________ 27. Serve as switches and amplifiers.

________ 28. Allow flow in only one direction.

________ 29. Create light from a filament heated by current flow.

________ 30. Low power, low cost, reliable lighting once only used for indicators.

A. CFL

B. Sensors

C. LEDs

D. Diodes

E. Transistors

F. Incandescent bulbs

G. Integrated circuits

Notes

Name ______________________________ Date ____________ Class ___________

Activity 8-1

Reading Analog Multimeters

Being able to read different types of meters is necessary for electrical engineers. Different types of meters include analog and digital meters, as well as multimeters capable of measuring multiple values. In this activity, you will learn to read analog multimeters.

Objectives

After completing this activity, you will be able to:

- Read an analog voltmeter and ammeter.
- Read an analog ohmmeter.

Materials

Pencil

Activity 8-1 Worksheet

Activity

In this activity, you will:

1. Use Activity 8-1 Worksheet to read ammeter and voltmeter scales and values.
2. Use Activity 8-1 Worksheet to read ohmmeter scales and values.
3. Complete the questions on Activity 8-1 Worksheet.

Reflective Questions

1. What role do you think the range selection plays in the accuracy of a measurement?

____

____

2. What do you think the advantage is of using an analog meter rather than a digital meter?

____

____

Activity 8-1 Worksheet

Voltage and current scales are linear, meaning that the numbers are evenly spaced. Notice below that the scale goes from 0 to 20. But it can actually read from 0 to 0.2 mA, 0 to 2 mA, 0 to 20 mA, 0 to 200 mA, or 0 to 2 A, depending on the range you select for your measurements.

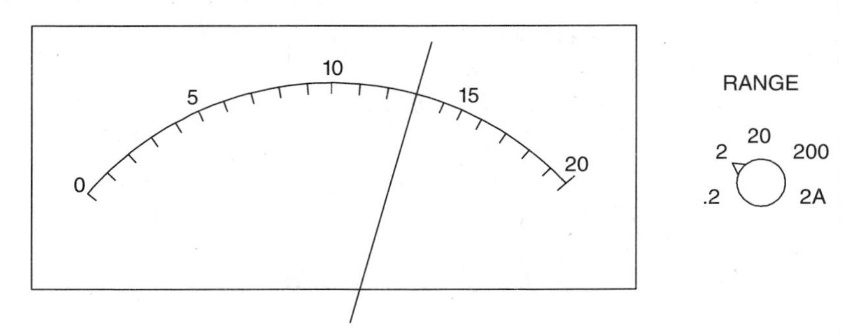

Look at the meters below and notice the range on which they are set. Read the meters and write your readings in the space provided.

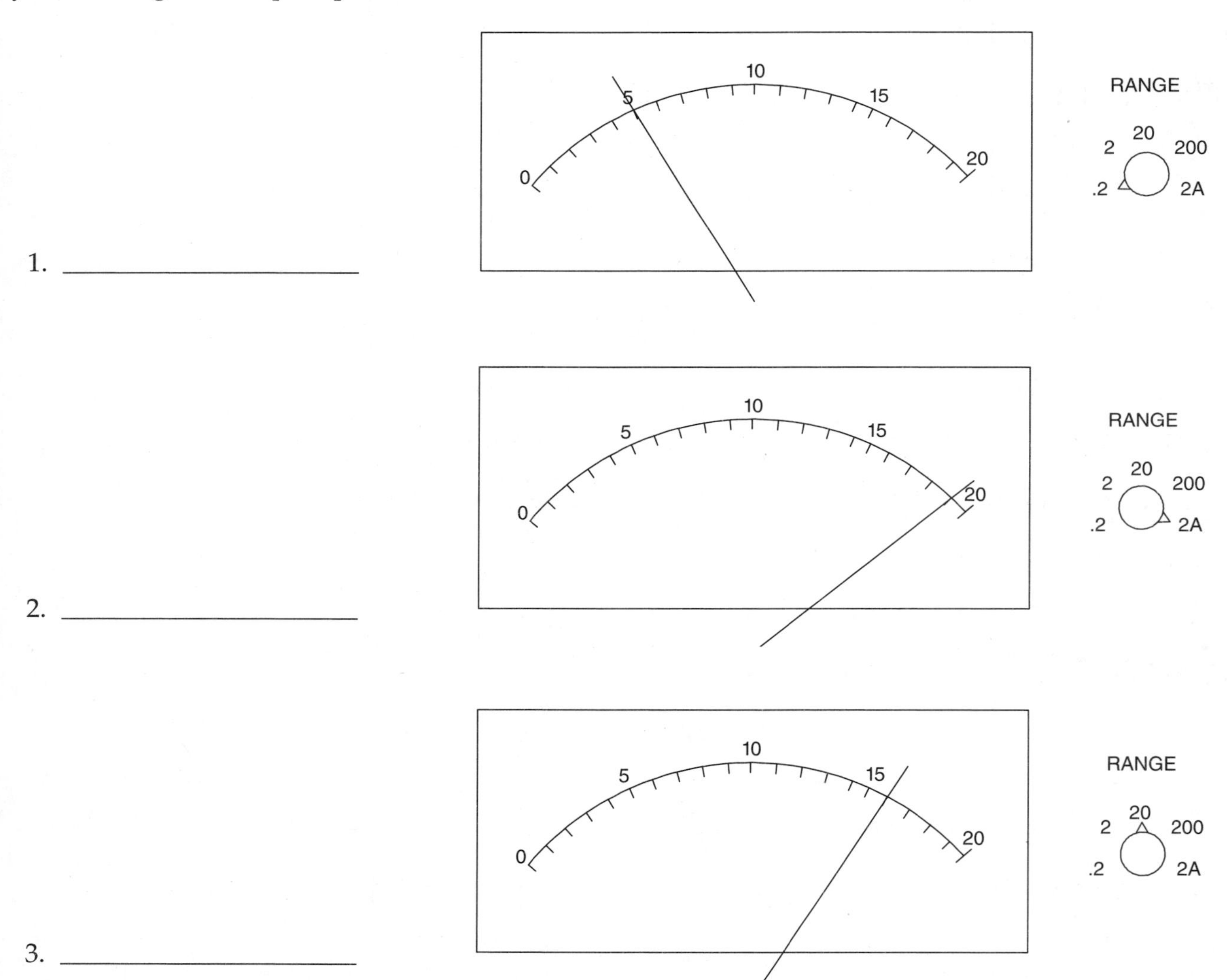

Activity 8-1 Worksheet *(Continued)* Name ____________________

Ohmmeter scales are nonlinear, which means that they are not evenly spaced. Notice how the spacing is much wider for the lower numbers and much closer for the higher numbers closest to infinite. Notice that the scale goes from 0 on the left to infinite on the right. Range switches are commonly used to aid in accurate measurement. For example, the meter could read 20 Ω. On the $\Omega \times 1$ scale, the reading would be 20 Ω. On the $\Omega \times 100$ scale, the reading would be 2,000 Ω.

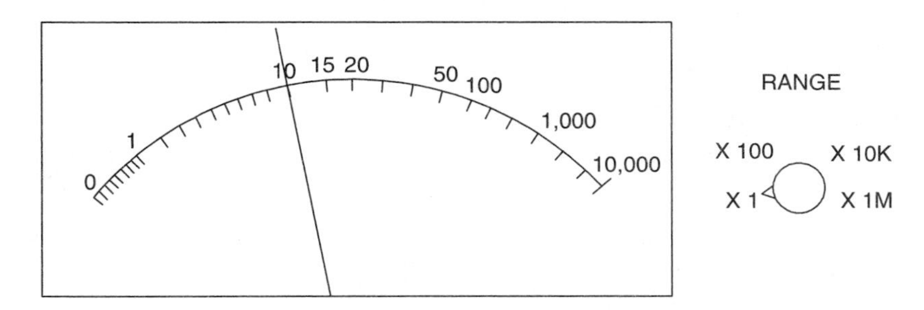

Look at the meters below and notice the range on which they are set. Read the meters and write your readings in the space provided.

0 1 5 10 20 50 100 1,000 10,000

RANGE

X 100 X 10K

X 1 X 1M

1. ____________________

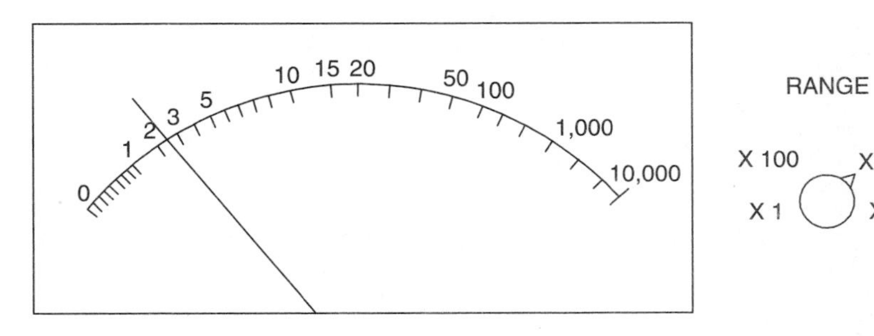

RANGE

X 100 X 10K

X 1 X 1M

2. ____________________

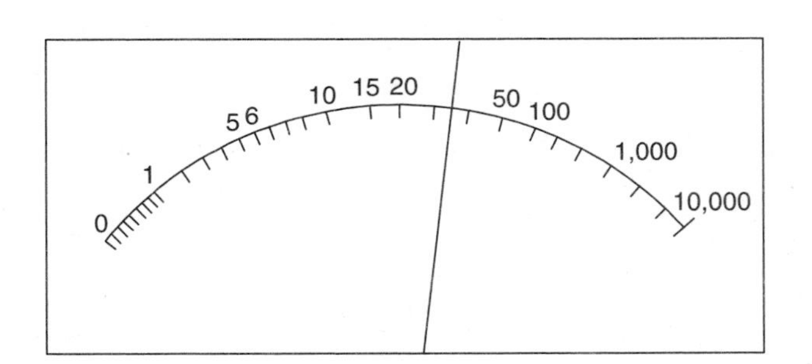

RANGE

X 100 X 10K

X 1 X 1M

3. ____________________

Notes

Name ______________________________ Date ____________ Class __________

Activity 8-2

Measuring Current, Voltage, and Resistance

To better understand how to read meters, it is essential that you learn to take measurements on real circuits. In this activity, you will build simple circuits and measure various values.

Objectives

After completing this activity, you will be able to:

- Build a simple circuit.
- Take measurements for voltage, current, and resistance.

Safety

Some measurements must be taken with the circuit energized, so it is extremely important to follow the lab safety rules outlined by your instructor to avoid injury to yourself and others.

Materials

Pencil

Activity 8-2 Worksheet

Circuit-building materials will vary depending on the teacher's discretion

Activity

In this activity, you will:

1. Build the circuits on Activity 8-2 Worksheet (or similar circuits based on the directions of your instructor) and take the measurements.
2. Remember that voltage measurements are always taken in parallel across a power supply or a load with the circuit energized.
3. Remember that current measurements are always taken in series with the circuit energized.
4. Always turn off the power and wire the meter into the circuit. Then, turn the power back on and take a measurement.
5. Take resistance measurements with the power off and the part(s) to be measured isolated from the rest of the circuit.

Reflective Questions

1. Why do you think the measurements are taken as they are, in series with, in parallel with, or isolated from the circuit?

2. Do you believe the measurements you took were accurate? Why or why not?

Name ______________________________

Activity 8-2 Worksheet

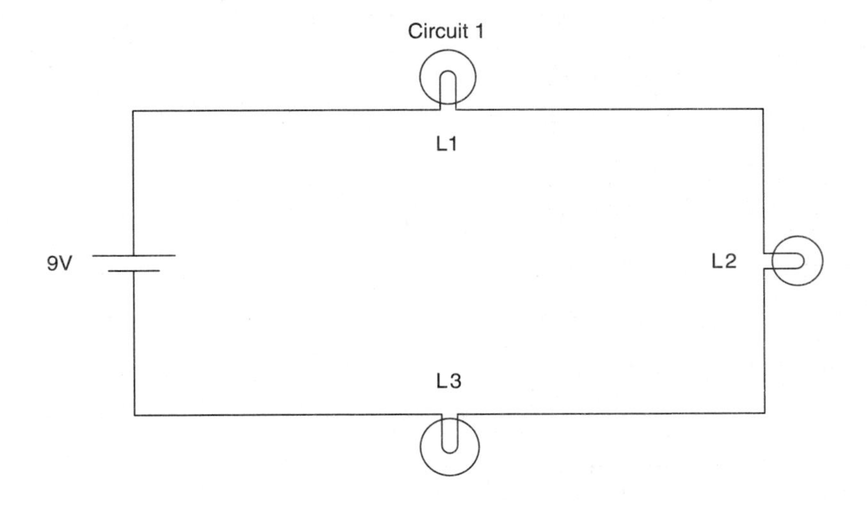

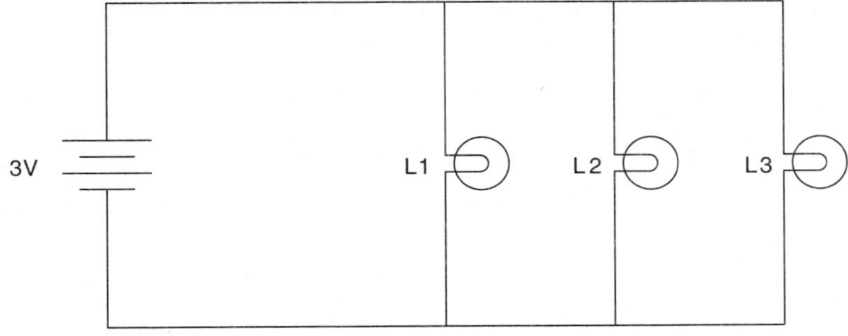

Notes

Name ______________________________ Date ____________ Class ___________

Activity 8-3

Electricity from Chemicals

Chemicals are used to produce electricity in the batteries and cells that start our cars and power our favorite portable electronic devices. A battery is made up of multiple cells connected together. A cell is simply made up of two unlike conductive materials called electrodes in an acidic electrolyte. One electrode becomes positive and the other becomes negative. When a load is applied to the electrodes, current will flow.

In this activity, you will use a grapefruit, copper nail, and galvanized nail to create a cell that will produce electricity. This activity can also be done using a penny and a nickel as electrodes and a piece of paper towel soaked in strong salt water as the electrolyte.

Objectives

After completing this activity, you will be able to:

- Make a simple cell using unlike conductive materials and an electrolyte.
- Test for voltage and current.
- Wire a circuit with a power source and a load.

Safety

Always wear safety glasses when working with chemicals. Wear the proper protective equipment and clothing when dealing with dangerous chemicals. Do not touch electrical circuitry when the power is on.

Materials

Copper nail (or any piece of copper)

Galvanized nail (or any piece of zinc)

Wire or alligator test leads

Small lightbulb

Grapefruit or other acidic fruit

Electrical meter

Activity

In this activity, you will:

1. Insert the copper and galvanized nails into the fruit about 2″ apart. Using the wire or test leads, connect the copper nail to one side of the light and the galvanized nail to the other side of the light. The light should turn on.
2. Measure and record the voltage across the lamp and the current flow in the circuit.

   ______________________ Voltage across the lamp

   ______________________ Current

3. Wire a second grapefruit cell in series with yours and observe the brightness of the lamp.
4. Measure and record the voltage across the lamp and the current flow in the circuit.

   ______________________ Voltage across the lamp

   ______________________ Current

Reflective Questions

1. What happened to the brightness of the lamp when the second cell was added?

   ____

   ____

   ____

2. What do you think would have happened if a stronger electrolyte had been used?

   ____

   ____

   ____

Name ________________________________ Date ______________ Class ____________

Activity 8-4

Electricity from Solar Cells

Solar cells produce electricity from light and can be used to power anything from a small calculator to large cities. Solar cells are a renewable green technology because they use light from the sun to produce electricity rather than burning fossil fuels.

Renewable energy is sure to be one of the fastest growing sectors of the economy for many years to come. Electrical engineers are on the forefront of design and implementation of these systems and must understand how they work.

In this activity, you will use solar cells to generate enough electricity to light a lightbulb and will measure the voltage and current in the circuit.

Objectives

After completing this activity, you will be able to:

- Use solar cells to power a small device.
- Show how multiple solar cells can be wired in series to increase output.
- Test for voltage and current.
- Wire a circuit with a power source and a load.

Safety

Always keep your hands away from electrical circuits when the power is turned on.

Materials

Solar cell

Lightbulb

Wire or alligator test leads

Electrical meter

Activity

In this activity, you will:

1. Using the wire or test leads, connect the solar cell to the light. The light should turn on.
2. Measure and record the voltage across the lamp and the current flow in the circuit.

    ______________________ Voltage across the lamp

    ______________________ Current

3. Wire two cells in series and observe the brightness of the lamp.
4. Measure and record the voltage across the light and the current flow in the circuit.

    ______________________ Voltage across the lamp

    ______________________ Current

5. Cover the cell with a solid object and observe the brightness of the lamp.

Reflective Questions

1. What happened to the brightness of the lamp when a second cell was wired in series?

    ____
    ____
    ____
    ____

2. How did your voltage and current readings change when the second cell was added?

    ____
    ____
    ____
    ____

3. What happened to the lamp when you covered the cell? Why?

    ____
    ____
    ____
    ____

4. How could solar cells be used to power electrical devices at night?

    ____
    ____
    ____
    ____

Name ______________________________ Date __________ Class __________

Activity 8-5

Electricity from Magnets

Electrical generators use magnets and motion to produce electricity. Most of the power that we use in our homes and schools comes from generators. Many different energy sources can be used to create the rotary motion needed for generators to produce electricity. Hydroelectric dams use the energy from falling water to turn generators. Wind turbines harness the power of the moving wind to turn their generators. Heat is created to create steam. The steam is then expanded across turbines to create rotary motion to turn generators.

Electrical engineers are heavily involved in the power generation field and must fully understand how electricity is produced from magnetism. In this activity, you will learn how magnets and motion can be used to generate electricity by passing a magnet through a coil of wire and testing the results on a meter.

Objective

After completing this activity, you will be able to:

- Demonstrate and explain how electricity is produced using magnets.

Materials

Bar magnet

Approximately 3′ of thin-gauge insulated wire

Electrical meter

Activity

In this activity, you will:

1. Make a coil of wire just large enough for the magnet to pass through with as many turns of wire as possible.
2. Set the meter to read current and connect both ends of the wire to the meter. Insert the magnet into the coil and then pull it back out.
3. You should see a small amount of current as you push the magnet into the coil and the same amount of current in the opposite polarity when you pull it back out.
4. You can see that current flows in a circuit when magnetic lines of flux are cut by a coil. This is the fundamental concept behind the operation of electrical generators.

Reflective Questions

1. Would current be generated if the magnet was held still and the coil was passed over it? Why or why not?

2. Explain how this concept can be applied on a larger scale to generate large amounts of electricity.

Name ______________________________ Date ____________ Class __________

Activity 8-6

Ohm's Law and Watt's Law

Electrical engineers use Ohm's law and Watt's law in the design and troubleshooting of electrical circuits. Ohm's law describes the relationship between current, voltage, and resistance. Watt's law describes the relationship between current, voltage, and power. In this activity, you will use Ohm's law and Watt's law to solve common circuit problems.

Ohm's law can be used to calculate current, voltage, or resistance when the other two values are known. Using the Ohm's law reminder, Figure 8-10 in the textbook, you can find the formula to solve each problem. Cover the quantity you are searching for with your finger, and you will see the formula. For example, cover the I with your finger, and you will see the formula E / R. This tells you that current can be found by dividing voltage by resistance.

Power (measured in watts) can be found using the "PIE" formula, where $P = I \times E$. By using the Ohm's law formula reminder with power, current, and voltage, you can find the necessary formulas.

Objectives

After completing this activity, you will be able to:

- Solve for current, voltage, or resistance when two of those values are known.
- Solve for power, voltage, or current when two of those values are known.
- Calculate the cost to operate a household device when its power usage and local billing rates are known.

Materials

Pencil

Textbook Figure 8-10

Activity

In this activity, you will:

1. Solve the following problems using Ohm's law:

_____________________ A. Calculate the voltage in a circuit with 50 Ω and 2 A.

_____________________ B. Calculate the current in a circuit with 50 V and 25 Ω.

_____________________ C. Calculate the resistance in a circuit with 120 V and 5 A.

_____________________ D. Calculate the resistance in a circuit with 10 V and 20 mA. To complete this problem, first convert 20 mA to amps.

2. Use "PIE" formula to solve the following problems:

_____________________ A. Calculate the power for a circuit with 100 V and 5 A.

_____________________ B. Calculate the current for a circuit with 1,000 W and 120 V.

_____________________ C. Calculate the voltage for a circuit with 500 W and 2 A.

3. Given the following information, solve the problems below.

Electrical billing is done in kilowatt-hours. The prefix *kilo-* stands for thousands, so billing is done in thousands of watt-hours. Let's say you turn on a 100-watt lightbulb for one hour. You use 100 watt-hours. If you leave it on for 10 hours, you use 1,000 watt-hours, or 1 kilowatt-hour, of electricity.

Imagine you have a small fan in your home that you like to have on at night, and you want to figure out how much it costs to operate that fan. The label says it draws 0.4 A at 120 V. First, you have to calculate the power used.

_____________________ A. Fan power

Assuming you operate your fan for 8 hours per night, 365 days per year, how many watt-hours are you using each year?

_____________________ B. Fan watt-hours

Assuming you are being billed at a rate of 15 cents per kilowatt-hour, what is the total cost to operate your fan every night for one year?

_____________________ C. Total cost to operate fan each year

Activity 8-6 *(Continued)* Name ______________________________

Reflective Questions

1. When do you think it might be more appropriate to calculate voltage, current, and resistance values rather than to test for them?

2. If voltage is held constant and resistance is increased, what will happen to current flow? Can you think of an example of where this is done in your home?

3. If people had a clear understanding of how much it costs per hour to operate each of their electrical devices, how do you think that would affect their behavior?

Notes

Name ________________________________ Date ____________ Class ____________

Activity 8-7

Circuit Identification

Series, parallel, and series-parallel circuits are wired in different ways. They each serve specific purposes. Electrical engineers need to understand these circuits and be able to tell them apart so they can work on them safely and effectively. In this activity, you will look at circuit schematics and determine if each is a series, parallel, or series-parallel circuit.

Objective

After completing this activity, you will be able to:

- Look at a circuit or circuit schematic and determine if it is a series, parallel, or series-parallel circuit.

Material

Pencil

Activity

In this activity, you will:

1. Look at the circuits below.

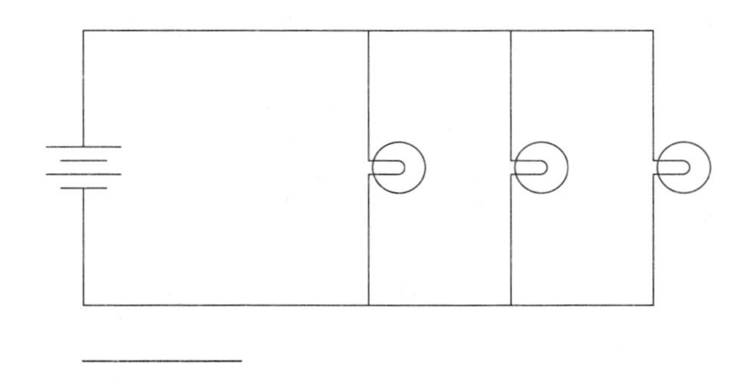

____

____

____

2. Determine if each circuit is:
   A. Series
   B. Parallel
   C. Series-parallel

Reflective Questions

1. Think of a practical example of where each of the following circuits is used:
   Series.

   Parallel.

   Series-parallel.

2. Are the circuits in your home wired in series or parallel? What would happen if the reverse were true?

3. If a string of lights were wired in series and one bulb burned out, what would happen to the rest of the lights? What if they were wired in parallel?

Name ______________________________ Date ____________ Class __________

Activity 8-8

Circuit Calculations

Electrical engineers must understand the specifics of series and parallel circuits in order to design and troubleshoot these circuits. Using some basic rules for series and parallel circuits in conjunction with Ohm's law, you can calculate information about these circuits. In this activity, you will solve a variety of problems in both series and parallel circuits.

Objectives

After completing this activity, you will be able to:

- Solve for current, voltage, and resistance in series and parallel circuits.
- Explain the rules for series and parallel circuits.

Materials

Pencil

Activity 8-8 Worksheet

Activity

In this activity, you will:

1. Recall the rules of a series circuit.
2. Solve for missing values of a series circuit on Activity 8-8 Worksheet.
3. Recall the rules of a parallel circuit.
4. Solve for missing values of a parallel circuit on Activity 8-8 Worksheet.

Reflective Question

1. Why do you think the rules of a series circuit are different from the rules of a parallel circuit?

____

____

Activity 8-8 Worksheet

In a series circuit, the following rules apply:

- Current is the same everywhere in the circuit.
- The sum of the voltage drops across each part equals the source voltage.
- The sum of the resistances of each load equals the total.

Look at the series circuit below and find the following:

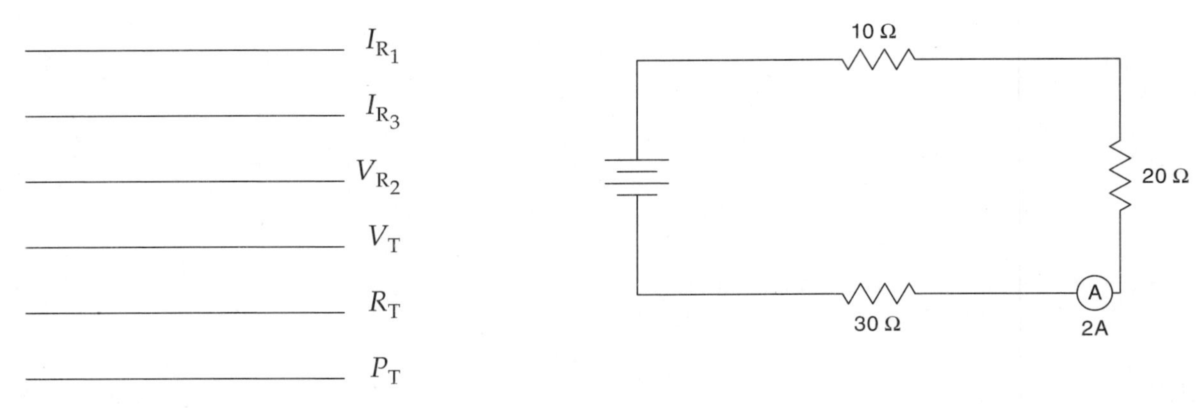

For parallel circuits, the following rules apply:

- The sum of the current in each parallel branch is equal to the total current.
- The voltage drop across each branch is equal to the source voltage.

Total resistance is found using the following formulas:

- With only two loads

$$R_T = \frac{R_1 \times R_2}{R_1 + R_2}$$

- With two or more loads

$$R_t = \frac{1}{\frac{1}{R_1} + \frac{1}{R_2} + \frac{1}{R_3} + \ldots + \frac{1}{R_N}}$$

Look at the parallel circuit below and find the following:

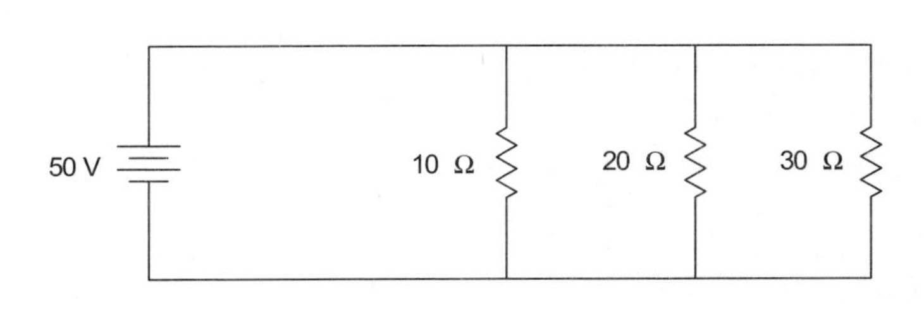

Name ______________________________ Date ______________ Class ____________

Activity 8-9

Resistor Color Code

Resistors are made with resistance values to meet specific requirements. The value of each resistor is indicated by color bands printed on the resistors. To work effectively in electronics, it is important to understand how to read these color bands. In this activity, you will be asked to find the value of various resistors.

Objectives

After completing this activity, you will be able to:

- Identify the value of a given resistor by looking at its color bands.
- Determine the highest and lowest possible measured readings for a given resistor based on its value and tolerance.

Materials

Pencil

Textbook Figure 8-15

Activity

In this activity, you will:

1. Use the resistor color code chart in Figure 8-15 of the textbook to find the values of the following resistors:

    ____________________ A. Brown, black, red

    ____________________ B. Blue, yellow, brown

    ____________________ C. Green, violet, orange

2. Use the following information to solve the problems below.

    On a four-band resistor, the fourth band is the tolerance. A gold-colored fourth band indicates a 5% tolerance, meaning that the actual measured resistance could be as much as 5% higher or lower than the indicated value. For example, a resistor with colors brown, black, brown, and gold would be a 100-Ω resistor with a 5% tolerance. Based on its tolerance, it could measure anywhere from 95 Ω to 105 Ω.

After finding the value and the tolerance, calculate the range of possible measured values for the following resistors.

_______________ A. Red, orange, brown, gold

_______________ B. Blue, violet, orange, silver

Reflective Question

1. What kinds of applications might require the tightest tolerances and the most accurate resistance values?

____

____

____

Name ______________________________ Date ____________ Class ____________

Activity 8-10

Continuity

Ohmmeters are commonly used in electronics to measure resistance, but they can also be used to test for continuity. A continuity test determines if there is a complete path for current to flow. A measurable amount of resistance indicates continuity. An infinite amount of resistance indicates a lack of continuity. For example, an ohmmeter can be used to check the continuity of a lightbulb to see if it has burned out. A reading of 100 Ω would tell you that there is continuity and the filament is intact. An infinite reading would indicate a lack of continuity and tell you that the filament had burned out. In this activity, you will troubleshoot a circuit using an ohmmeter.

Objective

After completing this activity, you will be able to:

- Use an ohmmeter to perform continuity tests.

Material

Circuit building materials, including four lamps

Ohmmeter

Pencil

Activity

In this activity, you will:

1. Wire four incandescent lamps in series.
2. Have another student unscrew one of the bulbs just enough that the connection is broken.
3. Use your ohmmeter to test for continuity across each lamp until you locate the one that has been unscrewed.

Reflective Questions

1. When might you need to perform a continuity test?

___

___

___

2. Would a reading of 500 Ω indicate continuity or a lack of continuity?

___

___

___

Name ______________________ Date ____________ Class ____________

Activity 8-11

Green Activity

In this activity, you will calculate the cost of an incandescent lamp compared with a compact fluorescent lamp. You will calculate how much money you can save by replacing the incandescent bulb with a CFL.

Objective

After completing this activity, you will be able to:

- Calculate the cost of an incandescent lamp compared with a compact fluorescent lamp.

Material

Pencil

Activity

In this activity, you will:

1. Imagine there is a 100-watt incandescent light inside or outside your home that your family leaves on for 10 hours each night.
2. Know that this lightbulb can be replaced with a compact florescent bulb that will create the same amount of light while using only 25 watts.
3. Calculate how much your family saves on your electric bill each year if you replace the incandescent bulb with a CFL at 15 cents per kilowatt-hour.

Reflective Questions

1. What might happen to your electric bill if you changed all of your bulbs to CFLs?

2. Do you think the same might be true about upgrading to more efficient appliances? Explain why or why not.

Name ______________________________ Date ______________ Class ____________

Activity 8-12

Project Construction

An electrical project can be soldered into a circuit board or connected using a breadboard. In this activity, you will build a project described by your teacher.

Objective

After completing this activity, you will be able to:

- Build an electrical circuit.

Materials

Circuit-building materials will vary depending on the teacher's discretion

Pencil

Activity

In this activity, you will be able to:

1. List each of the parts below and describe its function in the circuit.

____

____

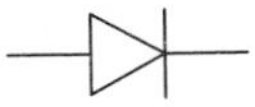

____

____

____

____

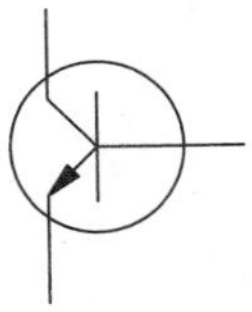

____

____

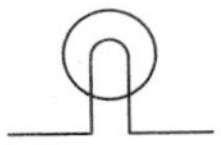

____

____

2. Build a circuit according to your teacher's directions.

Reflective Question

1. Why is it important to recognize different types of electrical symbols?

____

____

Chapter 9 Review

Civil Engineering

Name ______________________________ Date ____________ Class ____________

1. List three types of projects that civil engineers would design.

____

____

2. What are three subfields of the civil engineering?

____

____

3. What type of degree do civil engineering technicians typically have?

____

______________________ 4. The forces a structure must withstand are known as _____.

______________________ 5. The weight of a building is a(n) _____ load.

6. Why do structures that are not considered to be in a state of equilibrium fail?

____

____

Matching

_______ 7. Compression force

_______ 8. Tension force

_______ 9. Shear force

_______ 10. Torsion force

A. Pulling force

B. Forces that act in opposite directions across a material

C. Turning force

D. Crushing force

11. Describe the difference between beams and columns.

____

____

12. Describe the difference between a strut and a tie.

____

____

_______________ 13. The _____ is the device that connects two or more structural members together.

_______________ 14. A typical structure that is used to support a roof is known as a(n) _____.

15. Determine if a truss with 8 joints and 15 members is stable. Explain why or why not.

____

_____ 16. Which of the following bridges is used to span short distances?

A. Beam.

B. Cantilever.

C. Suspension.

D. Arch.

_____ 17. Which of the following bridges is used in the world's longest bridges?

A. Beam.

B. Cantilever.

C. Suspension.

D. Arch.

_______________ 18. A reinforced concrete shaft at the center of a skyscraper is known as a(n) _____.

_______________ 19. Land surveying is a branch of _____ engineering.

_______________ 20. The largest employer of civil engineers is the _____.

Name ______________________________ Date ____________ Class __________

Activity 9-1

Structural Forces

Understanding structures and structural forces is fundamental knowledge for civil engineers. Civil engineers use this knowledge to design safe and efficient buildings and structures. In this activity, you will find examples of different types of structures and structural forces.

Objective

After completing this activity, you will be able to:

- Provide examples of structural loads and structural forces.

Safety

Safety requirements will depend on the nature of the materials used in this activity and will be outlined by your teacher.

Materials

Poster board

Magazines

Newspapers

Computer with Internet access and printer

Scissors

Glue sticks

Activity

In this activity you will:

1. Find images (in magazines, newspapers, or on the Internet) that illustrate static and dynamic loads on a structure.
2. Label and glue the images to a poster board.
3. Find images that illustrate structural forces (compression, tension, shear, bending, and torsion). First, attempt to find images that represent these forces in structures. If you cannot find structural images, find images of materials under these forces, such as a rope under tension.
4. Label and glue the force images to the other side of the poster board.
5. Present your findings to the rest of the class.

Reflective Questions

1. How are dynamic and static loads different?

2. Why is it important for a civil engineer to understand structural forces?

Name ______________________________________ Date _____________ Class ___________

Activity 9-2

Structural Analysis

Civil engineers use structural analysis to ensure they are building structures that can withstand the necessary forces, without overbuilding the structure and wasting materials. In this activity, you will analyze a roof truss, which is one of the most fundamental structures.

Objectives

After completing this activity, you will be able to:

- Design a stable roof truss.
- Analyze a roof truss.

Materials

Truss design or structural analysis software (free/trial software is available online, such as Bridge Designer)

Activity

In this activity, you will:

1. Research truss designs and select a type of truss that you would like to design, such as Howe, Pratt, Gambrell, Fan, or Fink.
2. On paper, design a stable truss using the formula in the textbook ($2j = m + 3$).
3. Using a truss design software program, lay out the truss by:
   A. Placing the nodes.
   B. Adding the members.
   C. Adding loads.
4. Allow the software to calculate the forces in the truss.
5. Change the variables by adding additional forces or changing the amount of force.
6. Observe the differences.

Reflective Questions

1. In your first truss, which members were in compression and which were in tension?

____

____

2. Did the compression or tension forces change as you changed the variables?

____

____

3. What changes occurred as you changed the forces?

____

____

Name ______________________________ Date ____________ Class __________

Activity 9-3

Bridge Design Components

The design and construction of bridges is a common application of civil engineering. In this activity, you will model and label the components of a bridge.

Objectives

After completing this activity, you will be able to:

- Identify common bridge types.
- Recognize the components of a bridge through different bridge designs.

Safety

Safety requirements will depend on the nature of the materials used in this activity and will be outlined by your teacher.

Materials

Newspapers

Magazines

Computer with Internet access

Materials will vary based on the bridges that are being modeled

Balsa wood or pine strips

Plywood

Wire

Cardboard

File folders

Glue

Activity

In this activity, you will:

1. Research bridge designs and select a type of bridge that you would like to model, such as beam, truss, cantilever, or suspension.
2. Find several images of the type of bridge that you would like to model to use as a reference.
3. Model the bridge using the materials provided by your teacher.
4. Label the main components of the bridge, such as abutments, piers, deck, truss, suspension cables, and anchorages.
5. Present your model to the rest of the class.

Reflective Questions

1. What did you learn about the type of bridge that you built as you modeled it?

 ____

 ____

2. What similarities did you find among the bridges in your class?

 ____

 ____

3. What differences did you find among the bridges in your class?

 ____

 ____

Name ______________________________ Date ____________ Class __________

Activity 9-4

Structural Design

Civil engineers must use the engineering design process in order to solve problems. In this activity, you will solve a civil engineering design problem using the engineering design process.

Objectives

After completing this activity, you will be able to:

- Identify a civil engineering problem.
- Complete a civil engineering design problem.

Safety

Safety requirements will depend on the nature of the materials used in this activity and will be outlined by your teacher.

Materials

Materials will vary based on the design problem that is being solved

Balsa wood or pine strips

Plywood

Wire

Cardboard

File folders

Glue

Activity

In this activity, you will:

1. Use the chapter to determine a problem a civil engineer would be asked to solve.
2. Utilize the steps of the engineering design process presented in Chapter 2 to solve the problem:
   A. Problem definition
   B. Idea generation
   C. Solution creation
   D. Testing/analysis
   E. Final solution or output
   F. Design improvement
3. Throughout the process, keep an engineering notebook to record your notes, drawings, and findings.
4. Present your problem and solution to the rest of the class.

Reflective Questions

1. What did you learn about civil engineering from conducting the design process?

____

____

2. What other civil engineering problems could be solved using an engineering design process?

____

____

Chapter 10 Review

Mechanical Engineering

Name ______________________________ Date ____________ Class ____________

______________________ 1. True or False? Mechanical engineering is the newest field of engineering.

________ 2. The push or pull on an object resulting from contact with another object is known as _____.

A. force

B. power

C. pressure

D. energy

3. Define the engineering concept of *work*.

____

4. Explain the difference between potential energy and kinetic energy.

____

____

5. Define *power system*.

____

____

______________________ 6. The number of times a machine or tool multiplies the input force to perform work is known as _____.

Matching

_______ 7. Lever

_______ 8. Inclined plane

_______ 9. Wheel and axle

_______ 10. Screw

_______ 11. Wedge

_______ 12. Pulley

A. Doorstop

B. Ramp

C. Gears

D. Baseball bat

E. Device used at the top of a flagpole to raise and lower a flag

F. Device used in clamps to adjust pressure

13. Describe the difference between the fluids used in hydraulic and pneumatic power systems.

____

____

14. Provide two examples of each of the following power system components:

A. Power source.

____

____

B. Transmission device.

____

____

C. Control device.

____

____

D. Output device.

____

____

_______ 15. Which of the following is measured by multiplying distance and force?

A. Pressure.

B. Torque.

C. Work.

D. Power.

Chapter 10 Review *(Continued)* Name ______________________________

_______ 16. Which of the following is measured by dividing the amount of work by the amount of time?

A. Pressure.

B. Power.

C. Gear ratio.

D. Torque.

_______________ 17. True or False? The gear ratio 4:1 means that the driver gear is 4 times larger than the driven gear.

_______________ 18. True or False? Most machines operate at 100% efficiency.

19. Explain the difference between a simple gear train and a compound gear train.

____

____

20. Provide two examples of uses of hydraulics and pneumatics.

____

____

Notes

Name ______________________ Date __________ Class __________

Activity 10-1

Types of Motion

Understanding the types of motion is fundamental for a mechanical engineer. Throughout your daily life, you interact with a number of objects in motion. In this activity, you will observe everyday objects and describe their motion.

Objectives

After completing this activity, you will be able to:

- Provide examples of the four types of motion described in the textbook.

Materials

Pencil

Activity 10-1 Worksheet

Activity

In this activity, you will:

1. Be aware of the objects in motion with which you interact throughout the day. Use Activity 10-1 Worksheet to list examples of objects within each type of motion.
2. List the power source of the objects on Activity 10-1 Worksheet.
3. Present your findings to the rest of the class.

Reflective Questions

1.Were you surprised at the amount of materials that you use on a daily basis? Why or why not?

2.What did you learn about materials types from this assignment?

Activity 10-3 Worksheet *(Continued)* Name ____________________

Activity 10-1 Worksheet

Motion		
Material	**Object**	**Power Source**
Example: Rotary	Electric toothbrush	Battery-powered motor
Linear		
Linear		
Linear		
Linear		
Linear		
Rotary		
Rotary		
Rotary		
Rotary		
Rotary		
Reciprocating		
Reciprocating		
Reciprocating		
Reciprocating		
Reciprocating		
Oscillating		
Oscillating		
Oscillating		
Oscillating		
Oscillating		

Notes

Name ______________________________ Date ____________ Class __________

Activity 10-2

Mechanical Power Systems

Mechanical engineers design and build mechanical power systems. In this activity, you will examine a power system to determine the components that make up the system.

Objective

After completing this activity, you will be able to:

- Analyze a power system to determine the components.

Materials

Magazines

Newspapers

Computer with Internet access

Presentation software

Activity

In this activity, you will:

1. Select a power system, such as a bicycle, cordless drill, electric vehicle, arm on a backhoe, or pneumatic nail gun.
2. Conduct research to determine the components that make up the system.
3. Organize the components in the following categories:
   A. Power system
   B. Transmission device(s)
   C. Control device(s)
   D. Output device(s)
4. Find images of each of the components.
5. Create a presentation that illustrates each component in the power system you selected.
6. Present your findings to the rest of the class.

Reflective Questions

1. What did you learn about power systems from this activity?

____

____

____

____

____

____

____

2. How was your power system similar to and different from the power systems your classmates selected?

____

____

____

____

____

____

____

Name ______________________________ Date ____________ Class __________

Activity 10-3

Mechanical Power Formula

Mechanical engineers must be familiar with and capable of performing several different types of calculations related to work, pressure, power, torque, and gear ratios. In this activity, you will use a number of formulas to calculate mechanical power principles.

Objective

After completing this activity, you will be able to:

- Perform calculations related to mechanical principles.

Materials

Pencil

Activity 10-3 Worksheet

Activity

In this activity, you will:

1. Calculate answers to the problems given on Activity 10-3 Worksheet.

Reflective Question

1. How could any of these calculations be used in your daily life?

____

____

Activity 10-3 Worksheet

Calculate the following:

Work ($w = f \times d$) or ($w = p \times \Delta V$)

_____ 1. An engine jack is used to lift a 200-lb engine 4′. How much work is performed?

_____ 2. A winch is being used to pull a 3500-lb vehicle 30′. How much work is being performed?

_____ 3. An engine has a piston with a surface area of 12.56 in^2 and can travel 3.48 inches. What is the potential change in volume of the cylinder, given that ΔV = area × length?

_____ 4. If the engine in problem 3 was a V8 (meaning it has a total of eight cylinders), what would the total ΔV be for the entire engine, rounded to the nearest whole number?

_____ 5. If the piston in problem 3 had an input of 50 psi, what would the total amount of work performed by one piston be?

_____ 6. Convert the answer in problem 5 from in-lb to ft-lb.

Pressure ($p = f / a$) or ($f = p \times a$)

_____ 7. In a hydraulic jack, you have a 1000 lb of force being applied to a 3″ piston with a surface area of 7.065 in^2. What is the fluid pressure within the jack?

_____ 8. As an airplane is flying at 30,000 feet, the outside air pressure is 4.3 psi. The airplane windows are 16″ × 20″. What is the force being applied to the window?

_____ 9. A hydraulic system exerts a pressure of 50psi on a cylinder. If the desired output force is 350 lbs. what diameter is required for the cylinder? (Round your answer to the nearest 1/8″.)

Power ($p = w / t$) and Horsepower (in units of ft, lb, and minute) (hp = $(f \times d) / (t \times 33{,}000)$)

_____ 10. If the engine jack in problem 1 accomplished the task in 4 seconds, what would amount of power be?

_____ 11. If you applied a force of 150 lb to push a car 30′ in 2 minutes, how much horsepower would have been used?

Activity 10-3 Worksheet *(Continued)* Name ______________________________

Torque ($t = f \times r$) or ($f = t / r$)

______________ 12. You have a lug nut that needs 80 ft-lb of torque. You have a 14″ ratchet. How much force would it require?

13. Is it better to use a 8″ ratchet or a 14″ ratchet to tighten the lug nut? Why?

____

______________ 14. You have a system with two gears. The driver gear has 48 teeth and the driven gear has 12. What is the gear ratio?

Notes

Name ______________________________ Date ____________ Class __________

Activity 10-4

Mechanical Design

Mechanical engineers must use the design process in order to solve mechanical engineering problems. In this activity you, will solve a mechanical engineering design problem using the engineering design process.

Objectives

After completing this activity, you will be able to:

- Identify mechanical engineering problems.
- Complete a mechanical engineering design problem.

Safety

Safety requirements will depend on the nature of the materials used in this activity and will be outlined by your teacher.

Materials

Materials will vary based on the design problem that is being solved

Balsa wood or pine strips

Plywood

Wire

Gears

Pulleys

Cardboard

File folders

Glue

Activity

In this activity, you will:

1. Use the chapter to determine a problem that a mechanical engineer would be asked to solve.
2. Utilize the steps of the engineering design process presented in Chapter 2 to solve the problem:
   A. Problem definition
   B. Idea generation
   C. Solution creation
   D. Testing/analysis
   E. Final solution or output
   F. Design improvement
3. Throughout the process, keep an engineering notebook to record your notes, drawings, and findings.
4. Present your problem and solution to the rest of the class.

Reflective Questions

1. What did you learn about mechanical engineering from conducting the design process?

____

____

____

____

____

____

2. What other mechanical engineering problems could be solved using an engineering design process?

____

____

____

____

____

Chapter 11 Review
Bioengineering

Name ______________________________ Date ____________ Class ____________

1. What is the difference between biotechnology and bioengineering?

2. Identify the five areas of biology.

____________________ 3. The largest part and center of a cell is the _____.

4. What are two examples of homeostasis?

5. How does photosynthesis work?

6. What are *transgenesis* and *cisgenesis*?

____

7. What is biomass? Give an example of how biomass is used.

____

8. What is the difference between fermentation and anaerobic digestion?

____

_______ 9. The most direct form of thermochemical conversion is _____.

A. compost

B. combustion

C. pyrolysis

D. gasification

10. What two areas of production are the focus of agricultural engineering?

____

_______ 11. _____ are chemicals sprayed by farmers to eliminate weeds that may damage a specific crop.

_______ 12. True or False? The FDA has stated that offspring from clones should not be in the food supply.

13. Give four examples of biological engineering with humans.

____

Chapter 11 Review *(Continued)* Name ______________________________

14. List and describe three medical technologies.

____
____
____
____
____

15. How is ethics related to biotechnologies?

____
____
____
____
____
____

Notes

Name ______________________________ Date ____________ Class ____________

Activity 11-1

Prosthetic Device

Bioengineers create products to help individuals adapt to the world around them. One common area of bioengineering is the development of prosthetic devices. In this activity, you will create a prosthetic device to perform a specific task.

Objectives

After completing this activity, you will be able to:

- Design and build a working prosthetic device.
- Test and evaluate a prosthetic device.

Materials

20 craft sticks

String

Hot glue

Duct tape

Two different size cups, such as 12 oz and 4 oz

Activity

In this activity, you will:

1. Develop a prosthetic device that will allow you to drink from two different cups of water.
2. You must be able to pick up and set down both cups of water.
3. Your device must secure to your arm between your wrist and elbow.
4. You cannot use your other hand when testing the device.
5. You must work through the engineering design process to complete the task.

Reflective Questions

1. What was the most difficult part of the challenge? Why?

2. Would you make changes to the design? If so, what changes?

Name ______________________________ Date ____________ Class __________

Activity 11-2

Vertical Farms

Agricultural and biological engineers work with crops to ensure a safe and efficient food supply. Farmland is limited in some areas of the world. Engineers are beginning to look at vertical farms as a solution. In this activity, you will design and build a vertical farm system.

Objectives

After completing this activity, you will be able to:

- Design and build a vertical farm system.
- Evaluate the effectiveness of a vertical farm.

Materials

Fountain pump (big enough to lift water over 3′)

Flexible tubing for fountain pump

Small grow light

Growing medium, such as clay pellets or grow pods

Disposable aluminum baking pans

Material for frame structure, such as wood 2×2s, dowel rods, or small pvc pipe

Basil seeds

Activity

In this activity, you will:

1. Imagine you have been hired by your city to develop a vertical farm. The vertical farm cannot be taller than 3′ and must fit in a 3′ × 3′ base. To complete the activity, use the following steps:
   A. Work through the engineering design process to find a solution to this challenge.
   B. Research current vertical farm systems.
   C. Create a system to circulate water through the farm.
   D. Work with your teacher to determine specific supplies available for your vertical farm.
2. Share your design with the class and make adjustments as needed.

Reflective Questions

1. What could you improve on the design?

2. What would you need to change if this was a full-scale device?

3. Would this device be useful in your community? Why or why not?

Name ______________________________ Date ____________ Class __________

Activity 11-3

Genetic Engineering Debate

Engineers develop solutions to problems. Engineers must also determine the appropriate and ethical use of technology. Genetic engineering of food is a current topic of interest with appropriate and ethical concerns. In this activity, you will debate the production of genetically engineered food.

Objectives

After completing this activity, you will be able to:

- Research and analyze multiple viewpoints of a topic.
- Develop debate skills.
- Debate a current topic.

Materials

Newspapers

Magazines

Computer with Internet access

Activity

In this activity, you will:

1. Debate whether or not genetically engineered foods should be produced and/or how they should be regulated. The class will be broken into two teams: one for genetically engineered foods and one opposed to genetically engineered foods.
2. One person from each team will fulfill the following roles:
   A. Lead debater: Provides the introductory debate points through the opening statement.
   B. Questioner: Cross-examines the opposition
   C. Summarizer: Provides the rebuttal to close the debate
3. All members of the team will respond to questions from the opposition and moderator.

4. The debate will follow the following format:
   A. Affirmative position debater presents constructive debate points. (6 minutes)
   B. Negative position questioner cross-examines affirmative points. (4 minutes)
   C. Negative position debater presents constructive debate points. (6 minutes)
   D. Affirmative position questioner cross-examines negative points. (4 minutes)
   E. Moderator asks questions of both sides. (6 minutes)
   F. Negative position offers rebuttal. (5 minutes)
   G. Affirmative position offers rebuttal. (5 minutes)

Reflective Questions

1. Do you believe in the position you were assigned?

____

____

____

____

____

2. What new information did you discover?

____

____

____

____

____

3. Did you change your view of the issue?

____

____

____

____

____

Chapter 12 Review

Computer Engineering

Name ______________________________ Date ____________ Class ____________

1. How does computer engineering relate to software engineering and electrical engineering?

____

____

____

____

_______ 2. Logic gates typically employ transistors and diodes, but can also use _____.

A. resistors

B. capacitors

C. relays

D. coils

_______ 3. Inputs of 1 and 0 to an AND gate will provide an output of _____.

_______ 4. Inputs of 1 and 0 to an OR gate will provide an output of _____.

_______ 5. An input of 1 to a NOT gate will provide an output of _____.

_______ 6. What is a structured system of storing data in a computer system?

A. Algorithm.

B. Database.

C. Logic gate.

D. Spreadsheet.

_______ 7. Step-by-step procedures for solving problems or completing tasks are called _____.

A. algorithms

B. databases

C. logic gates

D. spreadsheets

_______ 8. _____ architecture refers to the way in which computers are designed.

A. Logic

B. Numerical

C. Computer

D. Software design

Matching

_______ 9. Integrated circuit that performs computer functions.

_______ 10. Main circuit board to which all other components connect.

_______ 11. Converts 120 volts AC to various DC voltages.

_______ 12. Stores data on rotating magnetic discs.

_______ 13. Allocate resources and organize and control hardware and software.

_______ 14. Memory that comes programmed on the device.

_______ 15. Temporarily stores data on which the computer is currently working.

_______ 16. Storage of frequently used data for easy access in the future.

A. Cache

B. Microprocessor

C. RAM

D. Power supply

E. Motherboard

F. ROM

G. Hard disc

H. Operating system

_______ 17. _____ is a manufacturing process where computers control the entire production process.

A. Computer numerical control (CNC)

B. Computer-integrated manufacturing (CIM)

C. Computer-aided design (CAD)

D. Robotics

_______ 18. _____ is the automation of machine tools using computers.

A. Robotics

B. Computer-aided design (CAD)

C. Computer-integrated manufacturing (CIM)

D. Computer numerical control (CNC)

Chapter 12 Review *(Continued)* Name ______________________________

_______ 19. The term _____ refers to the number of joints in a robotic arm.

A. Axis of movement

B. Degrees of freedom

C. Range of motion

D. Ergonomics

_______ 20. _____ is the name given to the sets of instructions for computers to follow.

A. Software

B. Hardware

C. Database

D. Microprocessor

Notes

Name ______________________________ Date ____________ Class __________

Activity 12-1

Logic Gates

Logic is a big part of computer design. Computer engineers must understand how logic gates work in order to design and troubleshoot electronic and computer circuitry. In this activity, you will identify and determine outputs of given logic gates.

Objectives

After completing this activity, you will be able to:

- Identify logic gates.
- Determine the outputs of given logic gates based on their inputs.

Materials

Pencil

Activity 12-1 Worksheet

Activity

In this activity, you will:

1. Match the logic gate symbols to the appropriate logic gate on Activity 12-1 Worksheet.
2. Give the correct output for the conditions of logic gates listed on Activity 12-1 Worksheet.
3. Draw a sample circuit for the given logic gates on Activity 12-1 Worksheet.

Reflective Questions

1. Name some logic gates used in the electronics you use every day.

____

____

____

2. Would the use of different logic gates be better in your electronics? Why or why not?

____

____

____

Notes

Name ____

Activity 12-1 Worksheet

Match the logic gates below with the symbol for that logic gate.

_______ 1. NOT

_______ 2. AND

_______ 3. OR

A

A
B
C

B

A
B
C

C

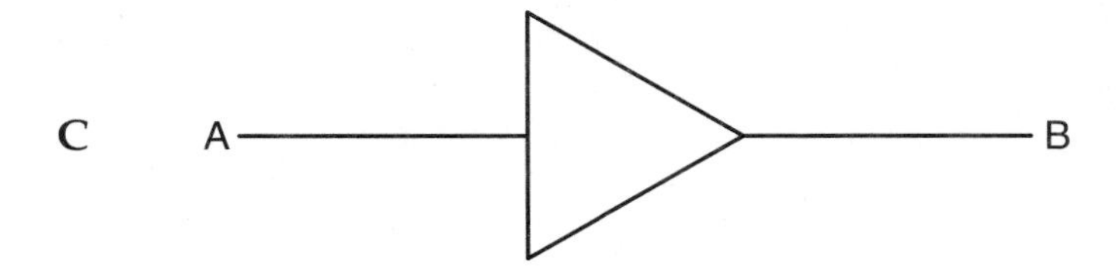

On the lines below, write the appropriate number to describe the output of the following.

_______ 1. AND gate with inputs of 1 and 0.

_______ 2. AND gate with inputs of 1 and 1.

_______ 3. AND gate with inputs of 0 and 0.

_______ 4. OR gate with inputs of 1 and 0.

_______ 5. OR gate with inputs of 1 and 1.

_______ 6. OR gate with inputs of 0 and 0.

_______ 7. NOT gate with an input of 1.

_______ 8. NOT gate with an input of 0.

In the space provided, draw a schematic for a circuit that simulates the action of each of the following gates.

1. AND

2. OR

3. NOT

Name ______________________________ Date ____________ Class __________

Activity 12-2

Algorithms

Algorithms are step-by-step procedures used for completing tasks or solving problems. In this activity, you will develop a problem-specific algorithm.

Objective

After completing this activity, you will be able to:

- Design an algorithm to solve a given problem.

Materials

Pencil

Activity 12-2 Worksheet

Activity

In this activity, you will:

1. Think of sorting through a list of numbers.
2. On Activity 12-2 Worksheet, write an algorithm that would find and identify the lowest number in the list.

Reflective Question

1. How could your algorithm be modified to solve other problems?

____

____

____

____

Activity 12-2 Worksheet

To find the area of a square, you might create an algorithm like this:

1. Ask for length of a side
2. Store length of side as s
3. Calculate area of the square (s × s)
4. Store the area of the square as a
5. Print a
6. Stop

In the area below, create an algorithm to find and identify the lowest number in a list of numbers.

Name ______________________________ Date ____________ Class __________

Activity 12-3

Computer Architecture

Computer architecture describes the design of computer hardware. In this activity, you will draw a diagram to better understand computer architecture.

Objectives

After completing this activity, you will be able to:

- Draw a block diagram showing how the main parts of a personal computer work together.
- Understand the relationship between major personal computer components.

Materials

Pencil

Activity 12-3 Worksheet

Activity

In this activity, you will:

1. Use Activity 12-3 Worksheet to design and draw a block diagram of the major parts of a personal computer, as discussed in the text.
2. Include the following components in your block diagram:
   A. Central processing unit (CPU)
   B. Microprocessor
   C. Motherboard
   D. Power supply
   E. Hard disc
   F. Monitor
   G. Keyboard
   H. Mouse
   I. Ports
   J. Speakers

Reflective Question

1. Can the location of the components used in computer architecture vary? Why or why not?

____

____

Name ______________________________

Activity 12-3 Worksheet

In the space provided below, draw a block diagram including all components used in computer architecture.

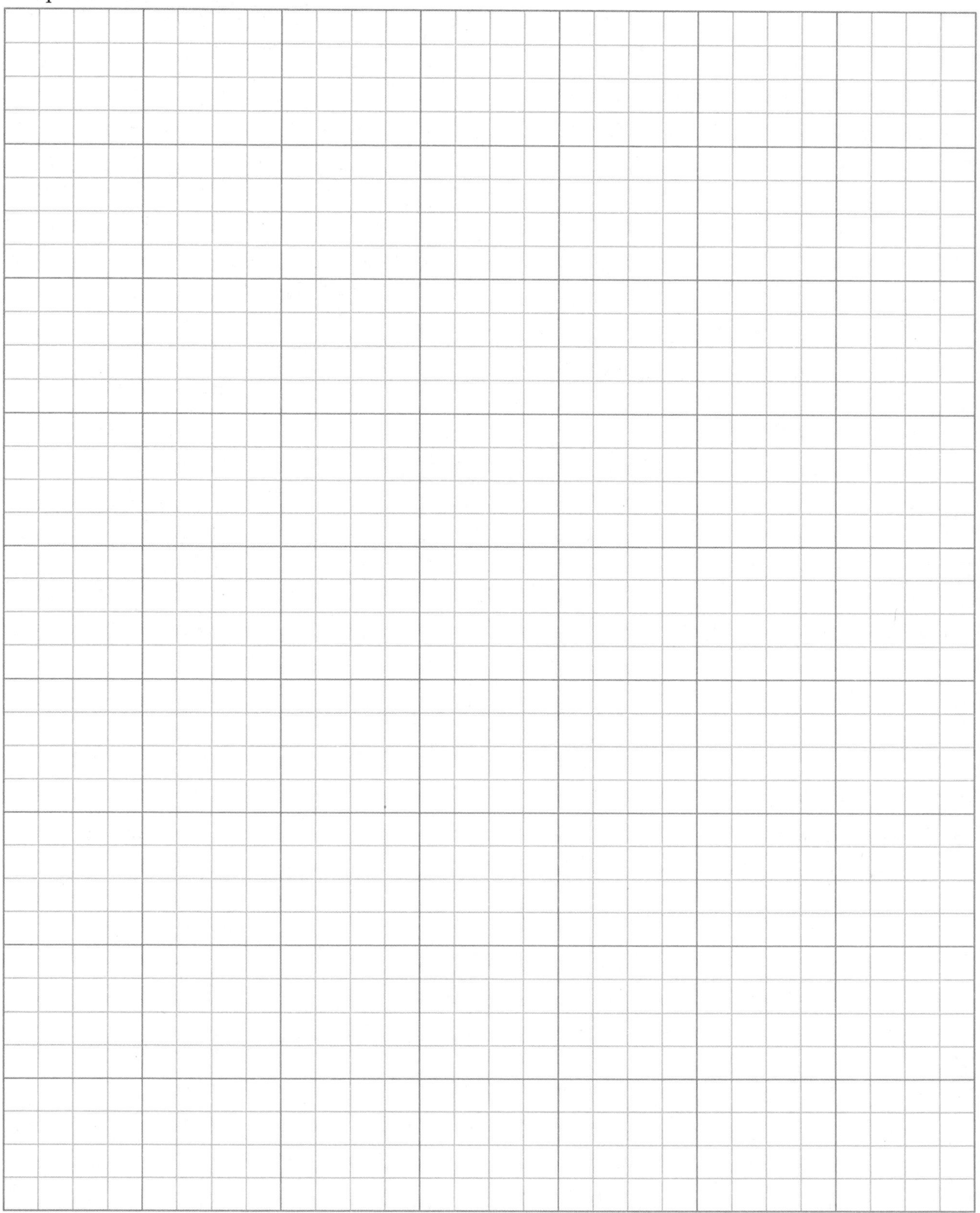

Additional graph paper

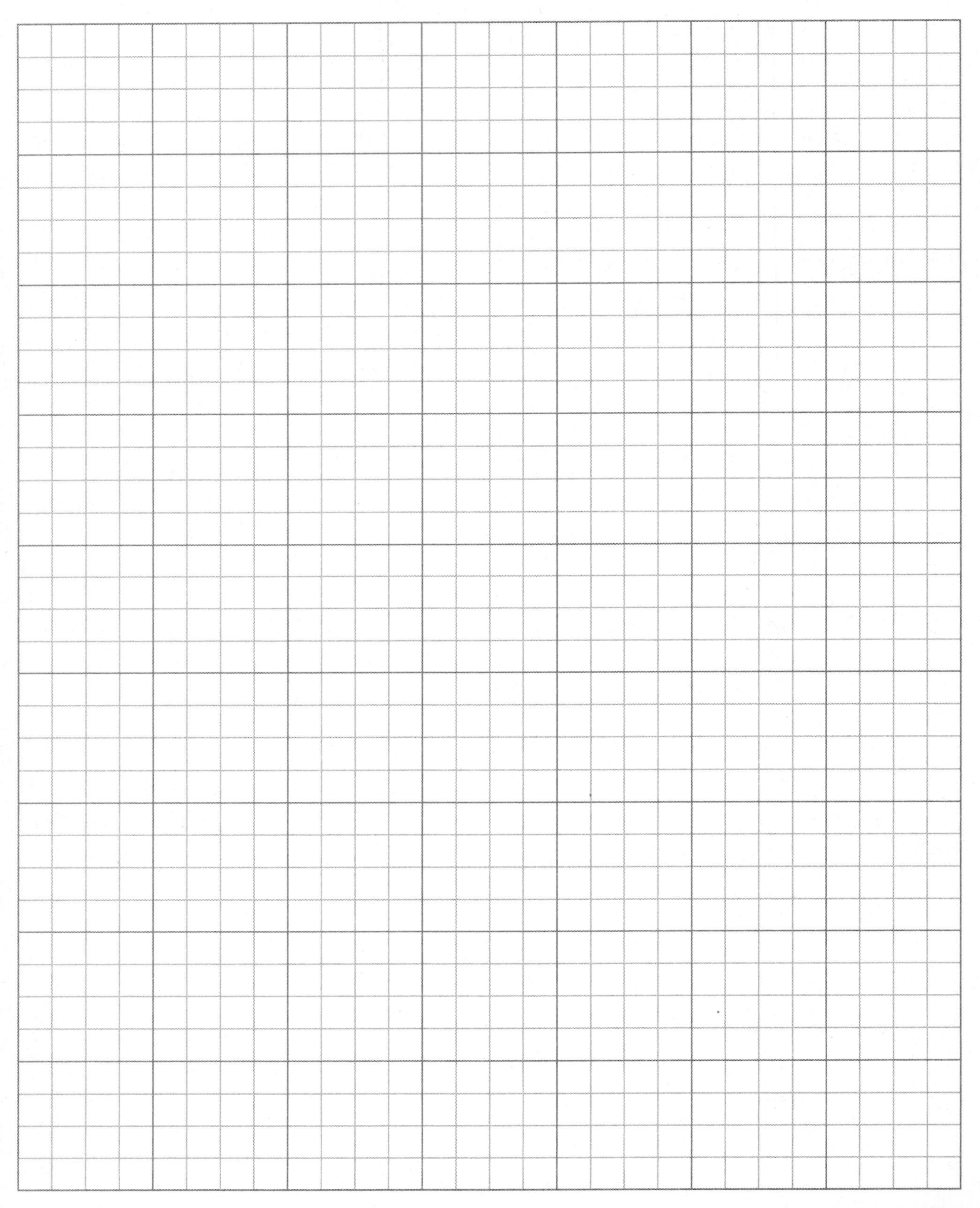

Name ______________________________ Date ____________ Class ____________

Activity 12-4

Binary Code

Digital signals are made up of series of 1s and 0s. 1s equate to an *on* condition and 0s equate to an *off* condition. Computers and other electronic devices use this system because transistors and diodes can react to the *on* or *off* conditions. In this activity, you will convert binary code to numbers and convert numbers to binary code.

Objective

After completing this activity, you will be able to:

- Convert numbers to binary code and binary code into numbers.

Materials

Pencil

Activity 12-4 Worksheet

Activity

In this activity, you will:

1. Convert numbers on Activity 12-4 Worksheet to binary code.
2. Convert binary code on Activity 12-4 Worksheet to numbers.

Reflective Question

1. What do you think is the relationship between binary code and logic gates?

____

____

____

Activity 12-4 Worksheet

Convert the following numbers in binary code.

______________________ 1. 29

______________________ 2. 35

______________________ 3. 99

______________________ 4. 133

______________________ 5. 159

Convert the following binary code to numbers.

______________________ 6. 10101010

______________________ 7. 10101

______________________ 8. 111000

______________________ 9. 100001

______________________ 10. 100100100

Name ______________________________ Date ____________ Class ___________

Activity 12-5

Building a Robotic Arm Using Syringes

Robots are automatically controlled, reprogrammable, multipurpose machines. Robotic arms are the most common form of robots used in manufacturing. In this activity, you will build a working robotic arm.

Objectives

After completing this activity, you will be able to:

- Design and build a robotic arm using syringes.
- Explain how compressed air can be used to move parts.
- Describe the movements of a common robotic arm.

Safety

You should receive training and approval from your teacher for each tool and piece of equipment you plan to use. Follow all safety rules set by your teacher. Make sure all hose connections are tight before compressing syringes.

Materials

Pencil

Activity 12-5 Worksheet

Newspapers

Magazines

Computer with Internet access

6 5-mL syringes

Wood for base, roughly 10″ square

1 4″ wooden disc

3 1/4″ × 4″ bolts with washers and nuts

6′ of 6 mm outside diameter hose to fit on syringes

2′ pieces of 3/4″ × 3/4″ wood

4 screw eyes to fit over the hose

Hot glue gun

Glue sticks

Small pieces of wood to make the arm and gripper (the gripper can be replaced with an electromagnet if small metal pieces are to be moved with the robotic arm)

2″ finish nails

Something to pick up and move, such as wooden blocks, balls, or toys

Wire ties

Activity

In this activity, you will:

1. Use the Internet, magazines, or other resources to research syringe robots and gather ideas for your design.
2. Sketch out your design on Activity 12-5 Worksheet, and have your design approved by your teacher before starting construction.
3. Design and build a robotic arm, which will use syringes for movement.
4. Use the robotic arm to pick up a series of items provided by your teacher and place them in a container in the least amount of time.

Reflective Questions

1. What do you think was good about your design?

___

___

___

2. What design improvements would you make to your robot?

___

___

___

Name ____

Activity 12-5 Worksheet

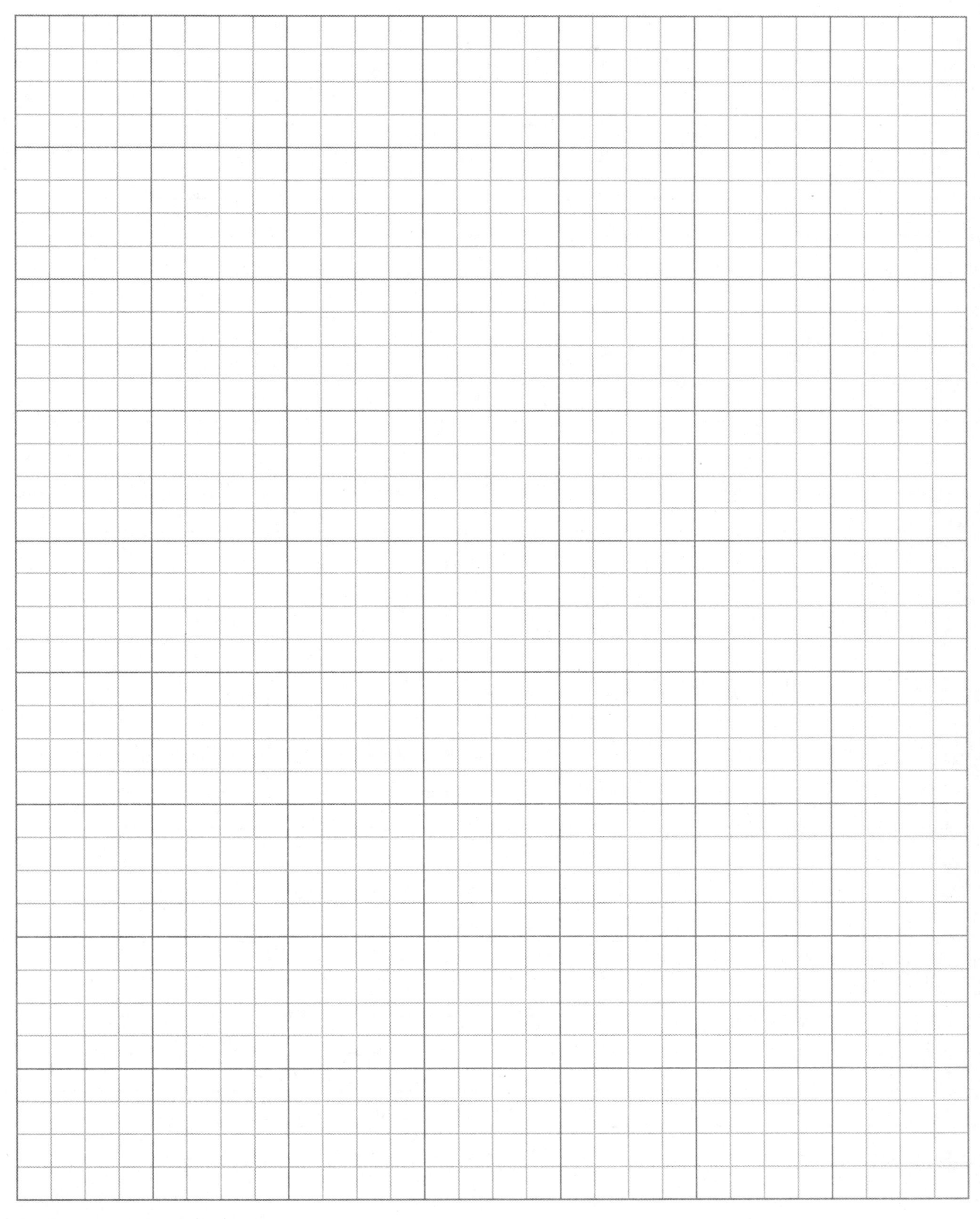

Notes

Chapter 13 Review

Aerospace Engineering

Name ______________________________ Date ____________ Class ____________

__________________ 1. _____ engineering deals with aircraft, and _____ engineering deals with spacecraft.

Matching

______ 2. Force is equal to the change in momentum per change in time

______ 3. For every action there is an equal and opposite reaction

______ 4. Every object persists in its state of rest or uniform motion in a straight line unless it is compelled to change that state by forces impressed on it

A. Newton's first law of motion

B. Newton's second law of motion

C. Newton's third law of motion

______ 5. _____ is the study of how air flows around solid objects.

A. Hydrodynamics

B. Aerodynamics

C. Aviation

D. Gravity

______ 6. The shape of a wing is called a(n) _____.

A. Design

B. Propeller

C. Airfoil

D. Lift

_______ 7. _____ is when the wing of a plane creates a downward force on the air, which in turn creates an upward force on the wing.

A. Drag

B. Propulsion

C. Lift

D. Gravity

_______ 8. Bernoulli's principle states that an increase in fluid speed creates a decrease in _____.

A. weight

B. pressure

C. mass

D. yaw

Matching

_______ 9. Force that moves aircraft through the air.

_______ 10. Force caused by gravitational pull of the earth.

_______ 11. Aerodynamic force that acts against movement of aircraft.

_______ 12. Upward force on the wing.

A. Lift

B. Thrust

C. Weight

D. Drag

_______ 13. _____ engines are used on extremely high-speed aircraft and spacecraft. Fuels are mixed and then explode in the combustion chamber.

A. Jet

B. Turbofan

C. Rocket

D. Turbojet

_______ 14. The *International Space Station* is an example of a _____.

A. satellite

B. lander

C. rover

D. flyby

(Continued) Name ____________________

Matching

______	15. Increase the amount of lift, especially at takeoff and landing.	A. Ailerons
______	16. Control up-and-down movement of the nose.	B. Rudder
______	17. Control side-to-side movement of the nose.	C. Elevator
______	18. Help the plane to turn.	D. Flaps

19. What are three ways to accommodate basic needs of the crew on a spacecraft in space?

____

____

____

____

____

____

____

____

____

____

____

____

20. Label the control surfaces on the plane below.

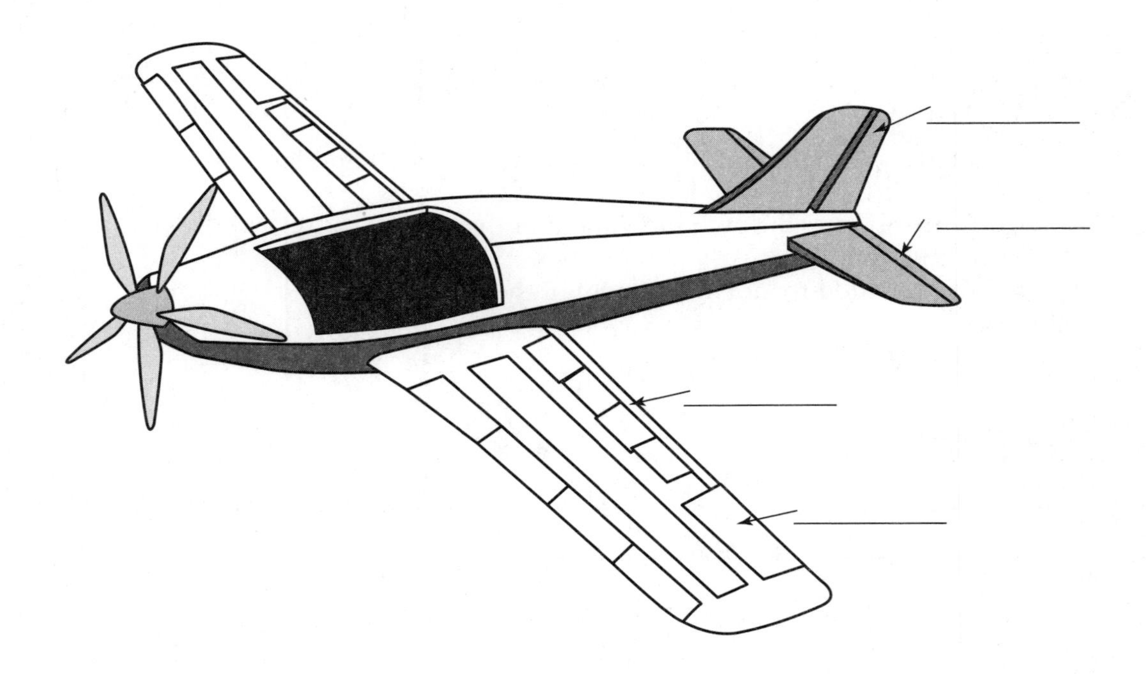

Name ______________________________ Date ____________ Class __________

Activity 13-1

Newton's First Law of Motion

Aerospace engineering relies on Newton's first law of motion to explain lift, thrust, and other concepts. A thorough understanding of this law is critical for your successful studies. In this activity, you will use a scale model of a car and a ramp to simulate a car crash and examine its effects on passengers with and without seat belts at various speeds.

Objectives

After completing this activity, you will be able to:

- Define Newton's first law of motion.
- Explain how Newton's first law of motion affects passengers in car crashes.

Safety

Make sure you are clear of the crash zone. There are sure to be flying materials depending on how fast your cars are moving at the point of impact.

Materials

Model car without a roof, such as a convertible, a truck, a modified or CO_2 car blank

Scale model of a passenger, such as a small toy person or something to simulate a person

Materials to build a ramp

Barrier for the bottom of the ramp

Tape measure

String

Activity

In this activity, you will:

1. Build a ramp for your car to accelerate down.
2. Place a barrier at the bottom that your car will crash into.

3. Place the passenger in the car and run the car down the ramp into the barrier.
4. Observe what happens.
5. Measure the distance the passenger travels out of the car.
6. Increase the height of the ramp and conduct the experiment again.
7. Measure the distance that the passenger travels.
8. Using the string, make a seat belt for the passenger and conduct the experiment again.

Reflective Questions

1. How far did the passenger travel beyond the barrier in the first crash?

____

____

2. How far did the passenger travel in the second crash with the steeper ramp?

____

____

3. Why was there a difference in travel distance when the ramp was higher?

____

____

____

____

4. How did your seat belt affect the passenger on impact?

____

____

____

____

Name ________________________________ Date ______________ Class ____________

Activity 13-2

Newton's Second Law of Motion

A force acting on an object will accelerate that object in the direction of the force. When you kick a soccer ball, it travels in the direction of the force you apply to the ball. Newton's second law of motion affects almost everything that moves. It provides a mathematical relationship between mass and acceleration. In this activity, you will observe this relationship by using a launching mechanism to accelerate balls of various weights.

Objectives

After completing this activity, you will be able to:

- Define Newton's second law of motion.
- Explain how the mass of an object affects its acceleration when force is constant.

Safety

You must remain clear of moving objects.

Materials

Balls of various weights, such as a baseball, tennis ball, ping pong ball, or shot put

Fixed anchor about 18–30″ from the floor, such as a desk or an easel

Roughly 3′ of string

Round disc of wood with a hole drilled in the center for the string, such as a 2×4 cut into a 3″ disc

Tape measure

Activity

In this activity, you will:

1. Feed the string through the hole in the wooden disc and tie a knot so the string cannot pull through.
2. Tie the other end of the string to your anchor point. Make sure the disc can swing freely in both directions to 90°.
3. Place the first ball directly in front of the wooden disc.

4. Pull the disc up to 90° and let go. The disc will swing down and make contact with the ball, causing it to accelerate across the floor.
5. Measure the distance it travels and repeat the process for the rest of the balls you chose.
6. You should find that distance traveled is inversely proportional to weight, so long as force is constant.

Reflective Questions

1. Which ball traveled farther? Why?

2. How can you use Newton's second law to explain your results?

3. Assuming constant force, what is the relationship between mass and acceleration?

Name ________________________________ Date ____________ Class ___________

Activity 13-3

Newton's Third Law of Motion

If you press your hands against your desk, you are exerting a force on the desk, but the desk is also exerting a force on your hands. Notice how your hands are slightly distorted and you can feel the pressure of the desk pushing against you. As you apply pressure on the desk, it applies the same pressure back on you.

Newton's third law of motion can also be demonstrated by taping a fan to a skateboard and turning on the fan. As the fan pushed air in one direction, an equal and opposite force is applied to the fan, and therefore the skateboard. The skateboard and fan will move across the floor. This works the same away as a rocket engine. In this activity, you will simulate a rocket using a balloon connected by a straw to a fishing line racetrack.

Objectives

After completing this activity, you will be able to:

- Define Newton's third law of motion.
- Use Newton's third law of motion to explain why a balloon moves in the opposite direction of the escaping air.

Safety

Be careful of parts flying around the room and balloons that explode.

Materials

About 30′ of fishing line

Drinking straws (one per team)

Balloons (one per team, plus some extras)

Tape

Activity

In this activity, you will:

1. Blow up your balloon, and tape a straw to the top of it.
2. While holding the balloon so no air can escape, pass the fishing line through the straw and hold the line tight.

3. Tie the other end of the line to a fixed point.
4. Let go of the balloon, and watch it travel up the string.

Reflective Questions

1. How can you use Newton's third law of motion to explain what happened to your balloon when you let it go?

___

___

___

2. What would have happened if you had a larger balloon or a balloon with greater pressure?

___

___

___

Name ______________________________ Date ____________ Class __________

Activity 13-4

Glider Design

Aircraft design depends on many variables. The first consideration to be made by the aerospace engineer is the intended use of the aircraft. Will it be used to haul cargo, move people, serve a military purpose, or perform some other function? How important is speed in relationship to fuel consumption? Once the needs are determined, the aircraft can be designed to meet those specific needs. Regardless of the purpose, aircraft are designed to be as light and fuel efficient as possible. They must create lift, have a suitable propulsion system, be easily maneuvered, and be safe. In this activity, you will design, build, and test a controllable balsa wood glider.

Objectives

After completing this activity, you will be able to:

- Design and built a controllable balsa wood glider.
- Explain how adjusting control surfaces changes the flight of a glider.
- Explain how changing the center of a glider's gravity affects its flight.

Safety

Only use cutting tools while in a seated position safely away from other people. Keep fingers away from cutting tools. Wear safety glasses. Wear carving gloves.

Materials

Patterns for cutting balsa and card stock

Card stock

Balsa wood

Artist knives

Sandpaper

Modeling glue

Large rubber bands

Adjustable launch ramp

Activity

In this activity, you will:

1. Build a balsa wood glider and glue on ailerons, an elevator, and a rudder for control of your glider.
2. Design the fuselage, wings, and horizontal stabilizer using two pieces of balsa wood glued flat together with a piece of card stock in the middle. The card stock should stick out an additional 1/4″ on the rear of each wing and on the rear of the horizontal and vertical stabilizers. These will be your control surfaces.
3. Keeping aerodynamics in mind, sand your glider until it is as smooth as possible and you have given the parts their proper shape.
4. Attach a small piece of modeling clay, if necessary, to balance your glider.
5. All gliders in your class will be launched from the same launcher to ensure uniformity of propulsion and launch angle. Each glider must have a hook on the bottom of the nose to attach the rubber band from the launcher.
6. When you are finished with assembly and balancing, it is time for your first test. Launch your glider from the launcher.
7. Retest your glider and make adjustments until you have the best possible setup for the glider that you made.

Reflective Questions

1. In your first test, did your glider nose down?

____

____

2. Did your glider turn left or right?

____

____

3. How could you make your glider fly farther?

____

____

____

4. If you were to build another glider, how would you change the design to achieve a longer flight?

____

____

____

Activity 13-4 *(Continued)* Name ______________________________

5. What effect did the following adjustments have on your flight?
   A. Ailerons

   ____

   ____

   B. Rudder

   ____

   ____

   C. Elevator

   ____

   ____

Notes

Name ______________________________ Date ____________ Class __________

Activity 13-5

Helium Dirigible

Dirigibles are lighter-than-air craft. This means they can fly because their weight is less than the weight of the air around them. Balloons are filled with hot air or gases, such as helium. You have probably seen hot air balloons in flight and blimps flying over major sporting events. In this activity, you will design and build a powered controllable dirigible with a gondola.

Objectives

After completing this activity, you will be able to:

- Describe the operation of lighter-than-air craft.
- Control the altitude, direction, and speed of a model dirigible.
- Safely work with tools and materials.

Safety

Only use cutting tools while in a seated position safely away from other people. Keep fingers away from cutting tools. Wear safety glasses. Wear carving gloves.

Materials

Helium tank

10-gallon trash bag

Light string

Balsa wood

Model cement

2 AAA batteries

1 holder for 2 AAA batteries

2 3-volt motors

Small plastic propellers to fit motors

Connecting wire

Clay for balancing weight

Activity

In this activity, you will:

1. Design a gondola that will be hung using strings under a 10-gallon trash bag full of helium.
2. Your gondola must accommodate a battery pack and two motors with propellers on them. You should design and build adjustable rudders to put behind the propellers for directional control. The helium provides a limited amount of lift, so keep the weight of your gondola to a minimum.
3. Your goal is to fly your dirigible from one side of the room to the other in a straight line without touching the floor or ceiling. Ideally, it should fly in a straight line without changing altitude.

Reflective Questions

1. Did your dirigible turn left or right during flight? If so, how can this be corrected?

____

____

____

2. Did your dirigible maintain the same altitude during flight? If not, how can this be corrected?

____

____

____

3. If you were to start over, how would you change your design?

____

____

____

____

____

Chapter 14 Review

Manufacturing Engineering

Name ______________________ Date __________ Class __________

_______________ 1. True or False? Manufacturing engineers must work in coordination with other engineers.

_______________ 2. _____ materials come from living matter.

_______________ 3. True or False? The Society of Automotive Engineers is the largest professional society for manufacturing engineers.

Matching

_______ 4. Organic materials mostly made up of carbon and hydrogen atoms.

_______ 5. Inorganic materials with good conductivity to heat and electricity.

_______ 6. Very hard, inorganic, refractory materials.

A. Metals

B. Ceramics

C. Polymers

Matching

_______ 7. Crude oil.

_______ 8. Trees.

_______ 9. Coal.

A. Drilling

B. Mining

C. Harvesting

_______ 10. Which method of tree harvest damages the environment the most?

A. Select cutting.

B. Seed tree cutting.

C. Clear-cutting.

D. Subsurface cutting.

11. Describe the difference between primary and secondary processing.

____

____

____

_______ 12. Scissors, saws, and drills are examples of common tools used in _____.

A. combining

B. separating

C. casting and molding

D. forming

13. What are the two types of bonding?

____

14. Describe each type of manufacturing that is used.

A. Continuous.

____

____

____

B. Intermittent.

____

____

____

C. Custom.

____

____

____

_______________ 15. _____ are used to guide tools to the correct location, and _____ are used to guide a work piece.

_______ 16. _____ are used to record and communicate the order of the manufacturing processes.

A. Flow process charts

B. Operation sheets

C. Operation process charts

D. Plant layout diagrams

Chapter 14 Review *(Continued)* Name ________________________________

_____________________ 17. When using _____ layout, the facility is designed around the manufacture of one product.

________ 18. Using a(n) _____ delivery system cuts the cost of warehousing large supply inventories.

A. commercial

B. just-in-time

C. mechanical

D. organized

19. Explain why quality control is important.

____

____

____

____

20. Explain how effective scheduling can keep production costs down.

____

____

____

Notes

Name ____________________ Date ____________ Class __________

Activity 14-1

Design Production

Once a design has been approved, it must be manufactured. In this activity, you will plan and set up a production line.

Objectives

After completing this activity, you will be able to:

- Create a plant layout.
- Create various process charts.
- Design tooling.
- Design a safety program.

Safety

Safety is the single most important consideration in any manufacturing activity. You should be trained and approved by your teacher before you use any tool or piece of equipment.

Materials

Pencil

Activity 14-1 Worksheets

Materials may vary depending on design

Activity

In this activity, you will:

1. Create a design to solve a problem or choose an existing design to be manufactured. Have your design approved by your teacher.
2. Using Activity 14-1 Worksheet A, create a plant layout drawing for your facility with arrows showing the path your parts will take through the facility.
3. Using Activity 14-1 Worksheet B, fill out a flow process chart for each part.

4. Draw an operation process chart showing each of your parts through the entire process on Activity 14-1 Worksheet C.

5. Fill in the operations process sheets for your parts using Activity 14-1 Worksheet D.

6. Design and draw each piece of tooling you plan to build and use for your production run on Activity 14-1 Worksheet E, and have them approved by your teacher.

7. Develop a safety program for your facility using Activity 14-1 Worksheet F. Your program should outline safety procedures to be followed by all team members, including protective clothing and goggles to be worn, training that will occur prior to the production run, and anything else you and your teacher agree is important. Have your safety program approved by your teacher.

8. Design a quality control program that will ensure your products will meet quality standards. Use Activity 14-1 Worksheet G to describe your quality control program and to draw any gauges you plan to use.

Reflective Questions

1. Did your facility run the way you expected? Why or why not?

____

____

____

____

2. Could you improve the design of your facility to make manufacturing tasks more efficient?

____

____

Name ____

Activity 14-1 Worksheet A

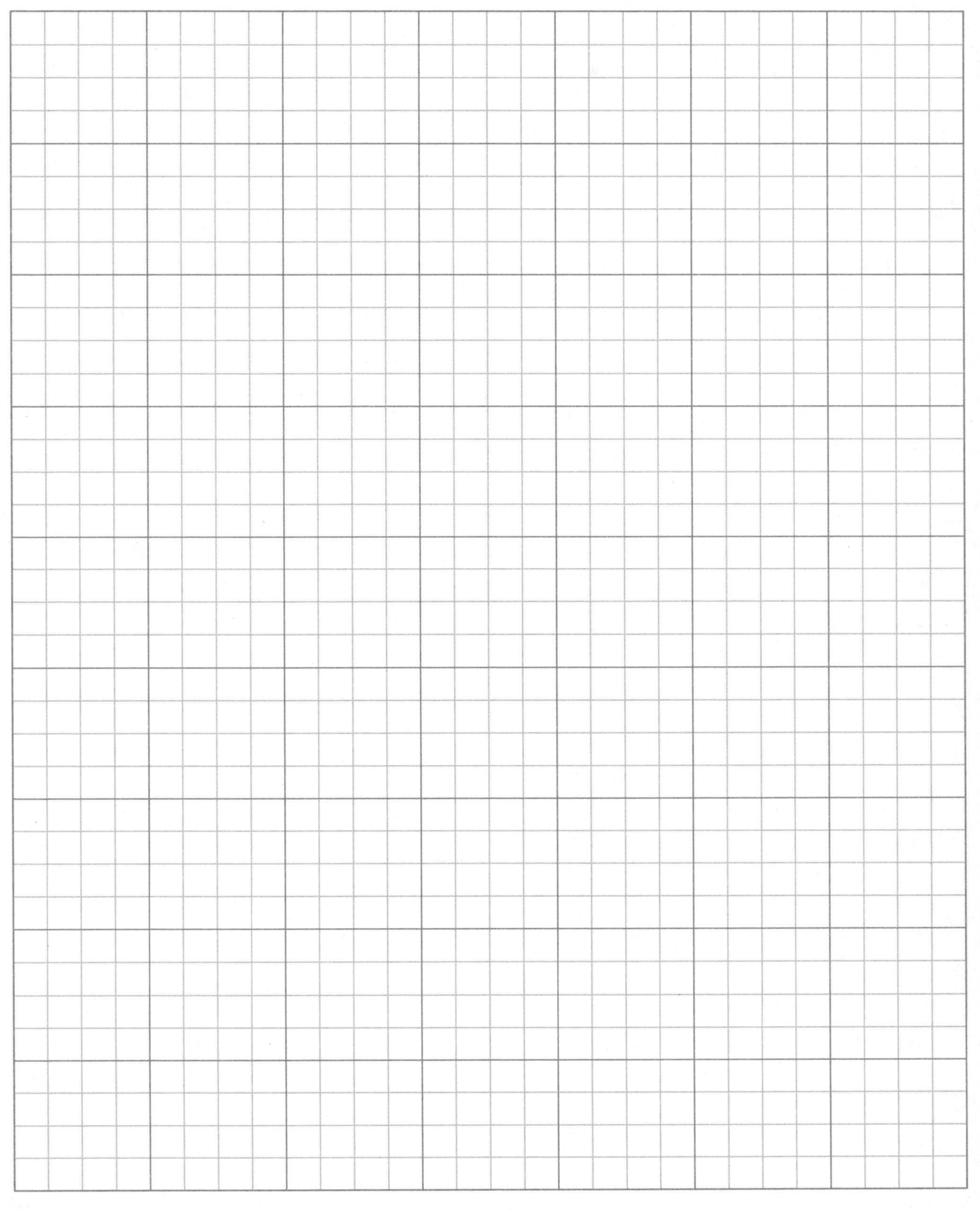

Activity 14-1 Worksheet B

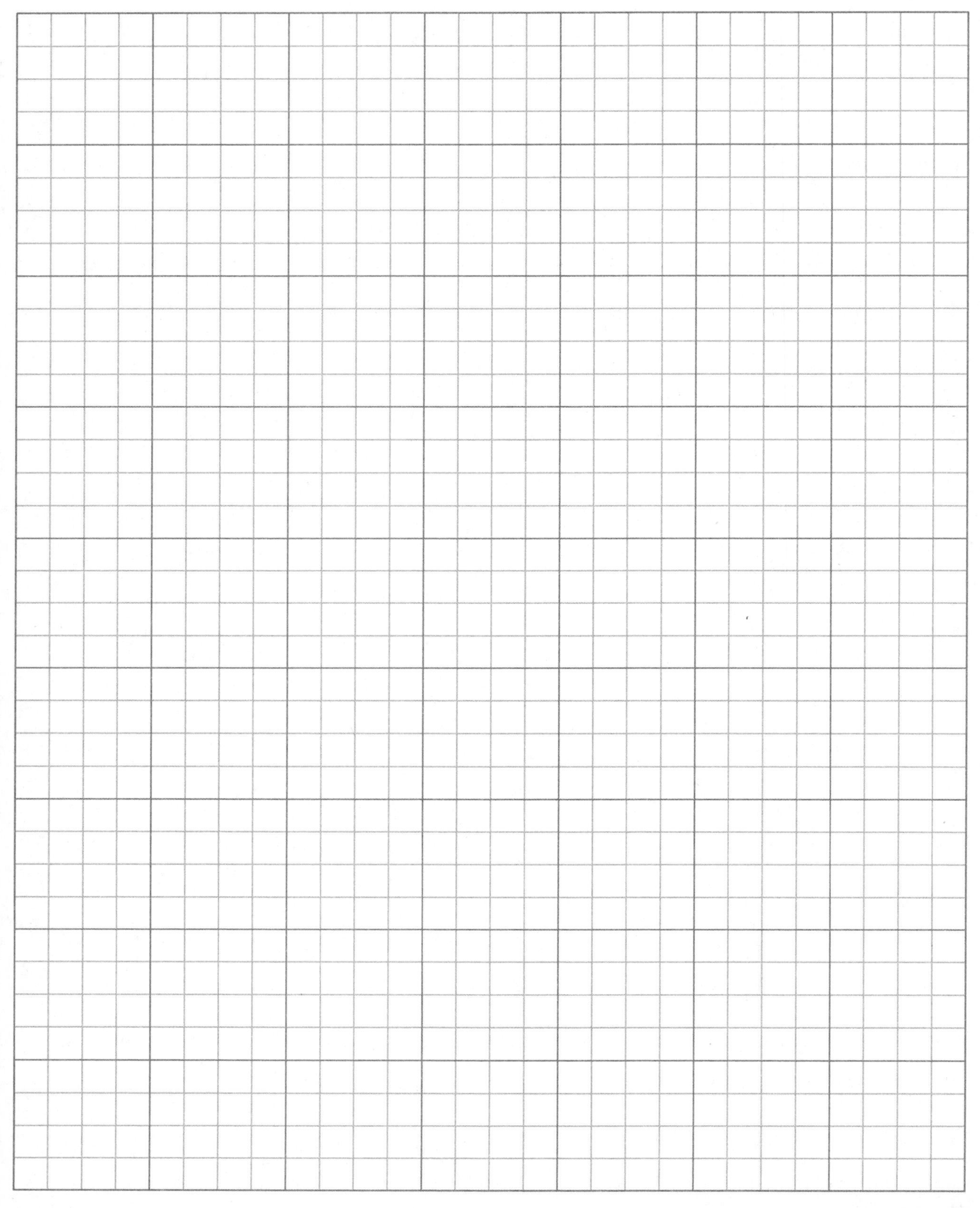

Activity 14-1 Worksheet B *(Continued)* Name ______________________________

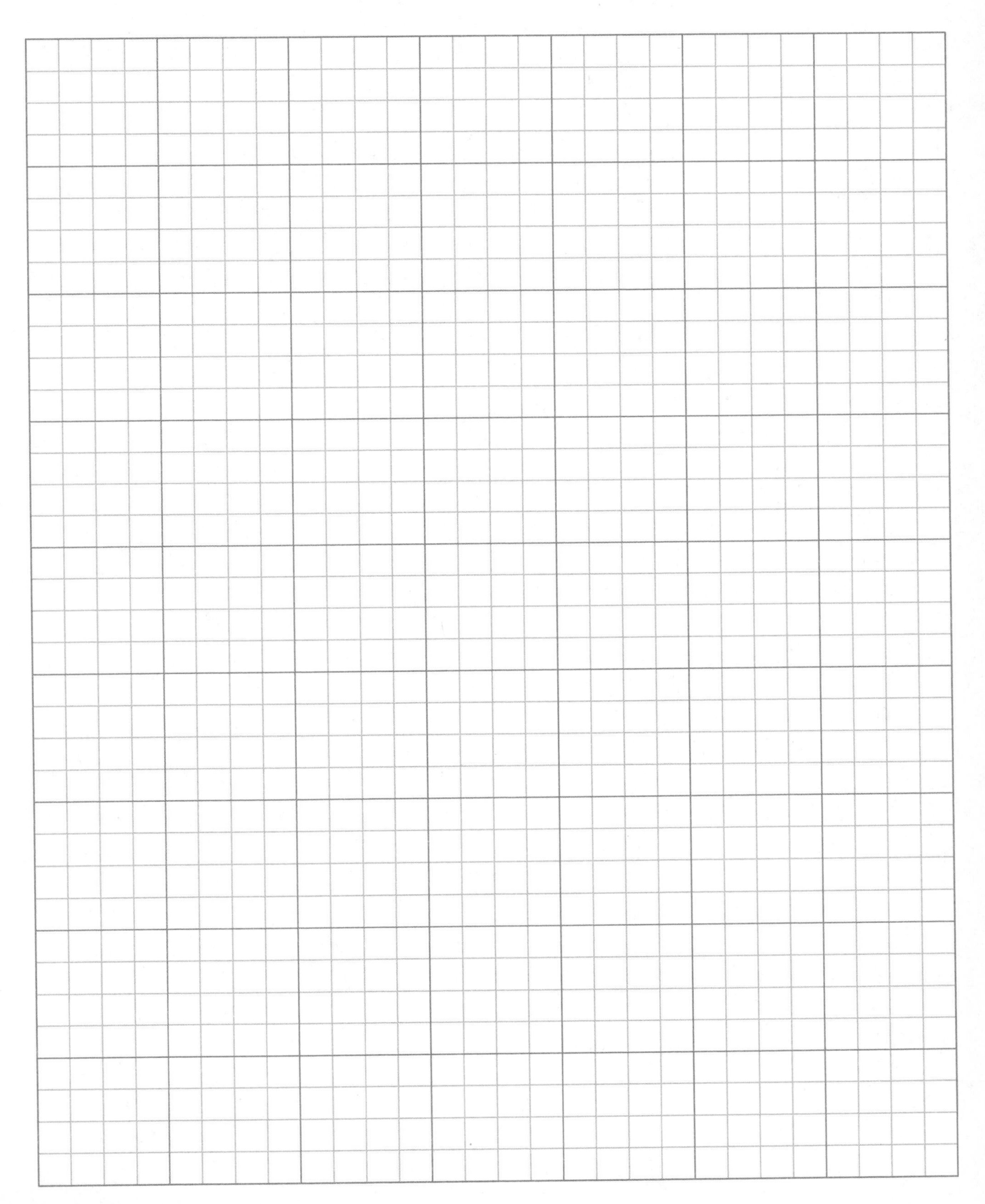

Additional graph paper

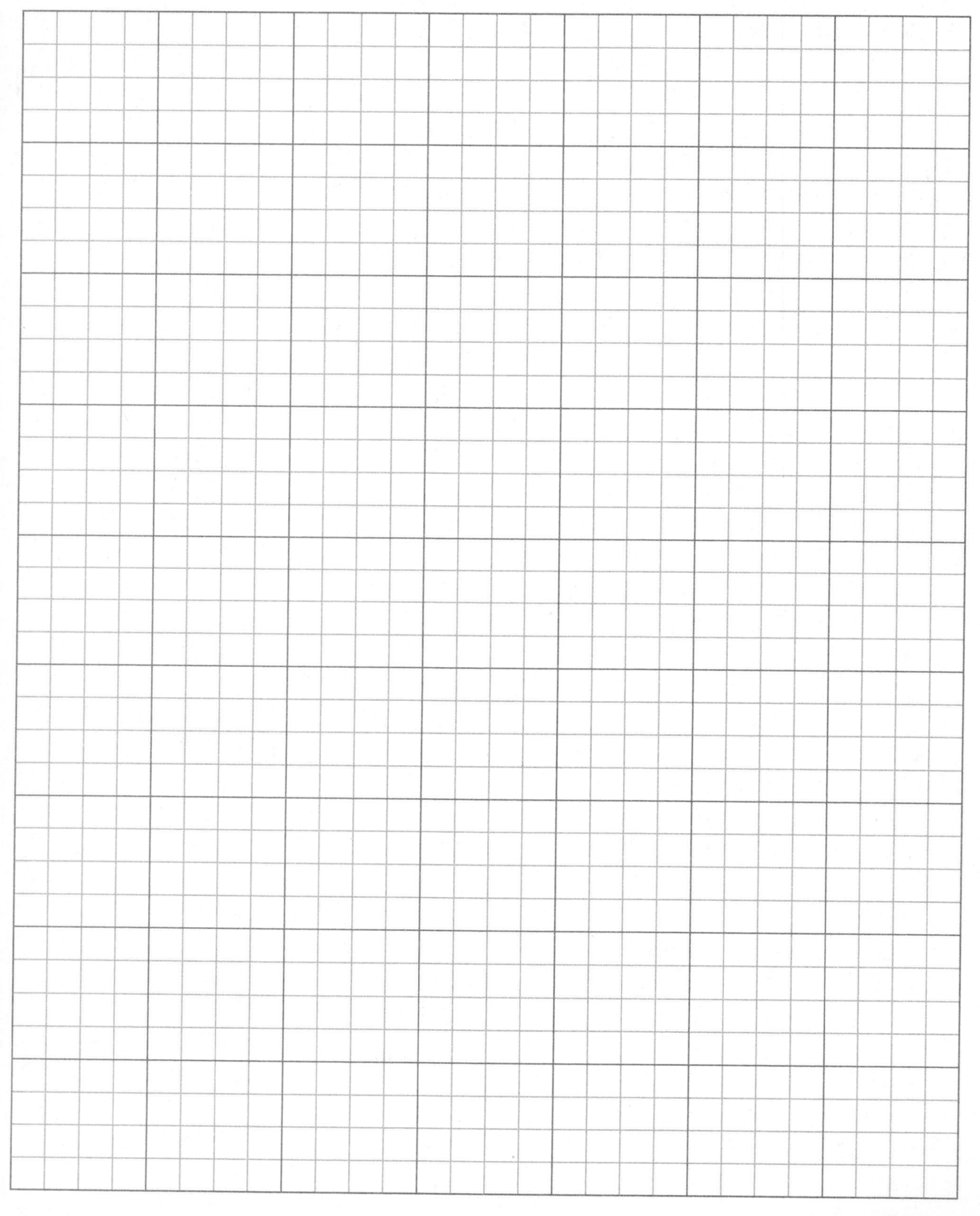

Name ____________________

Activity 14-1 Worksheet C

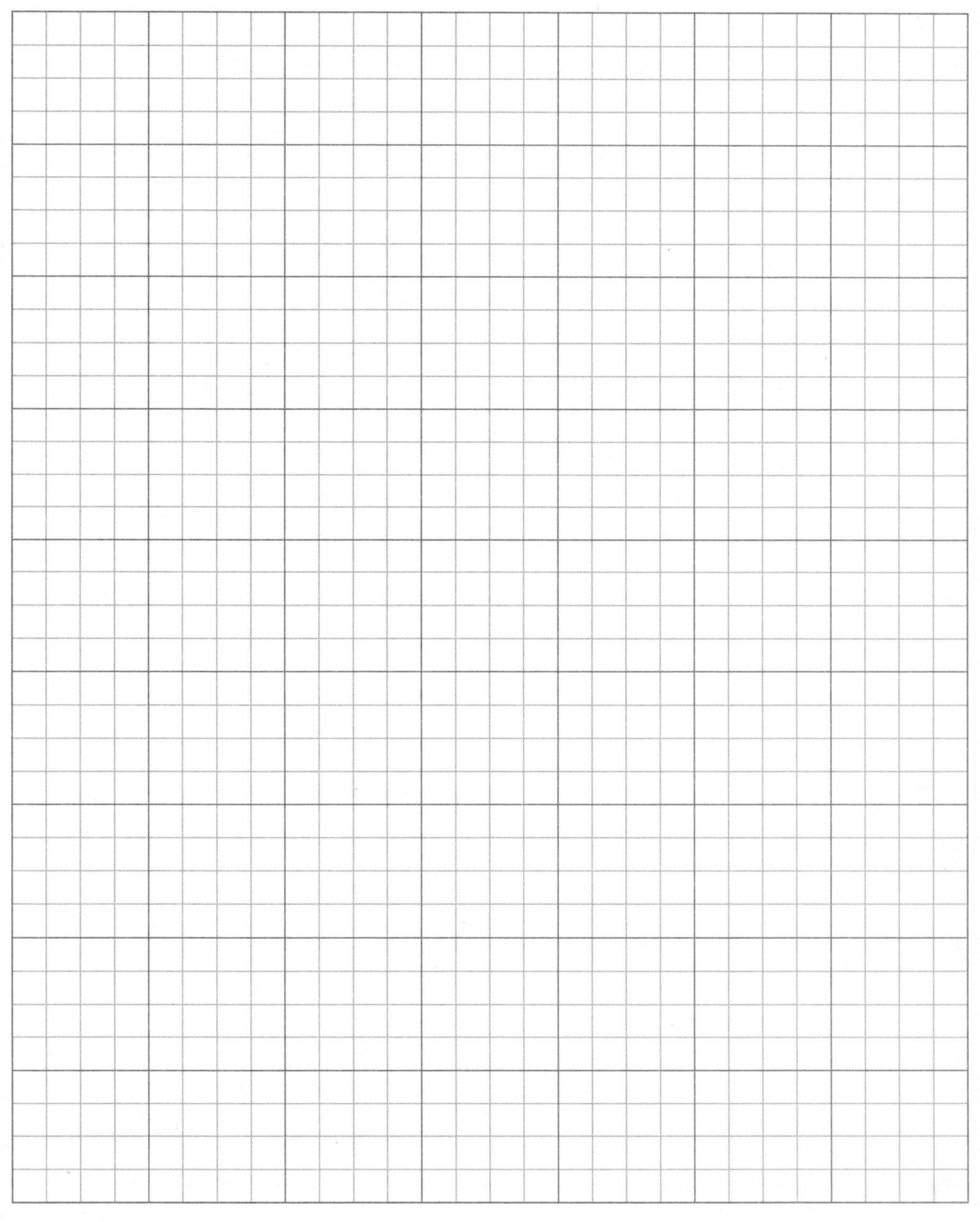

Activity 14-1 Worksheet D

Activity 14-1 Worksheet D *(Continued)* Name ____________________

Activity 14-1 Worksheet D *(Continued)*

Name ____

Activity 14-1 Worksheet E

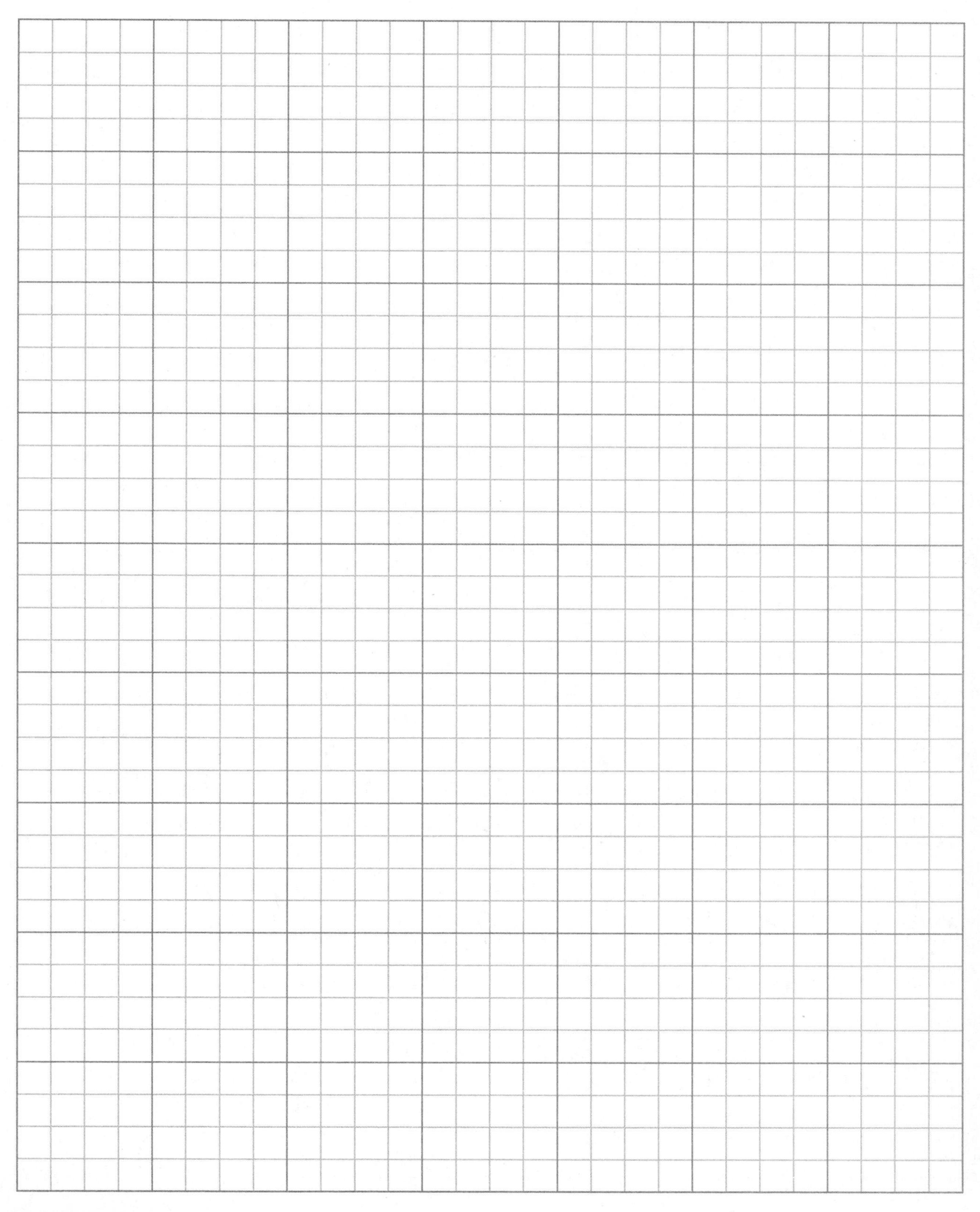

Activity 14-1 Worksheet E *(Continued)*

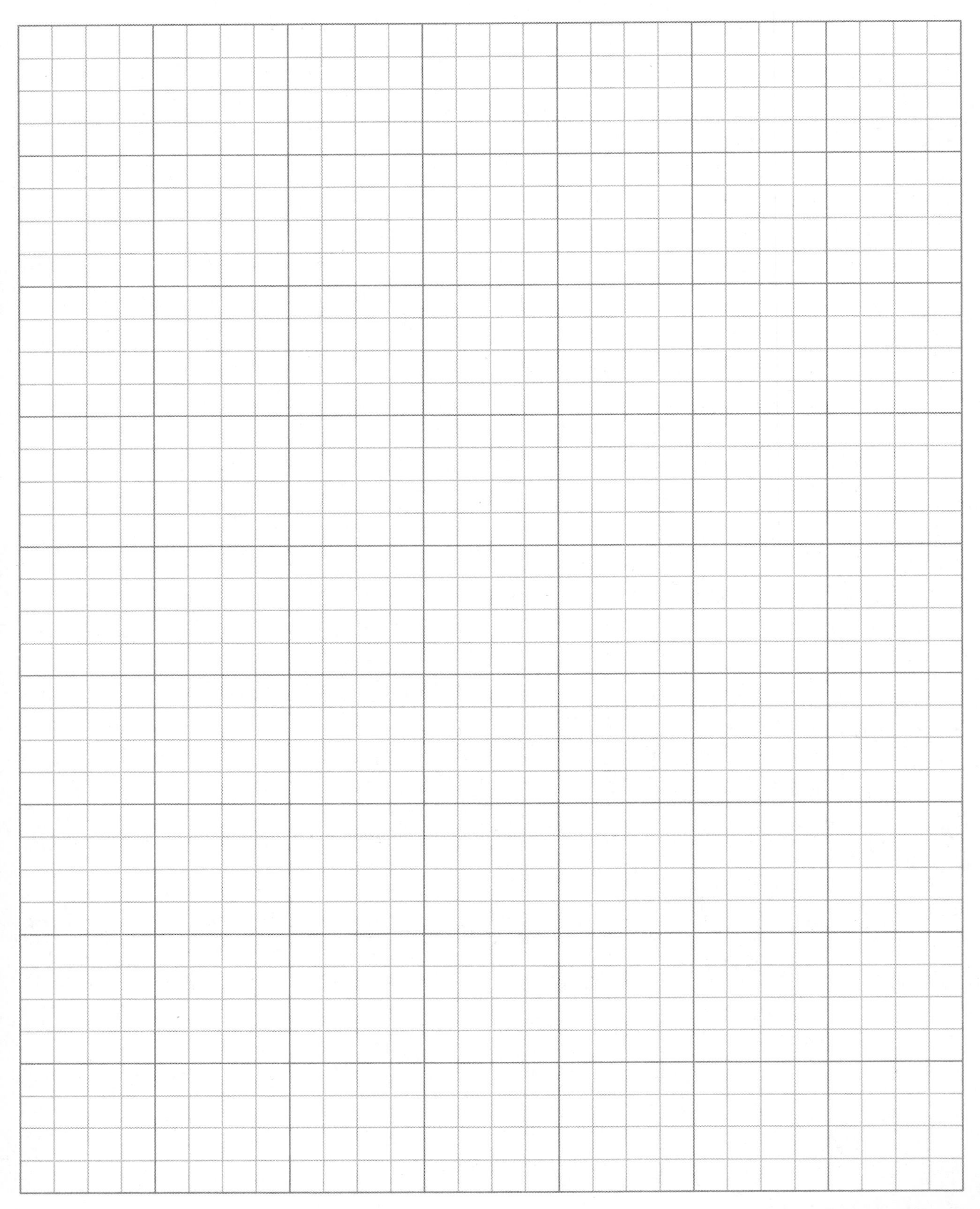

Name ______________________________

Activity 14-1 Worksheet F

Activity 14-1 Worksheet G

Activity 14-1 Worksheet G *(Continued)* Name ______________________________

Draw any quality control gauges you will create and use during production.

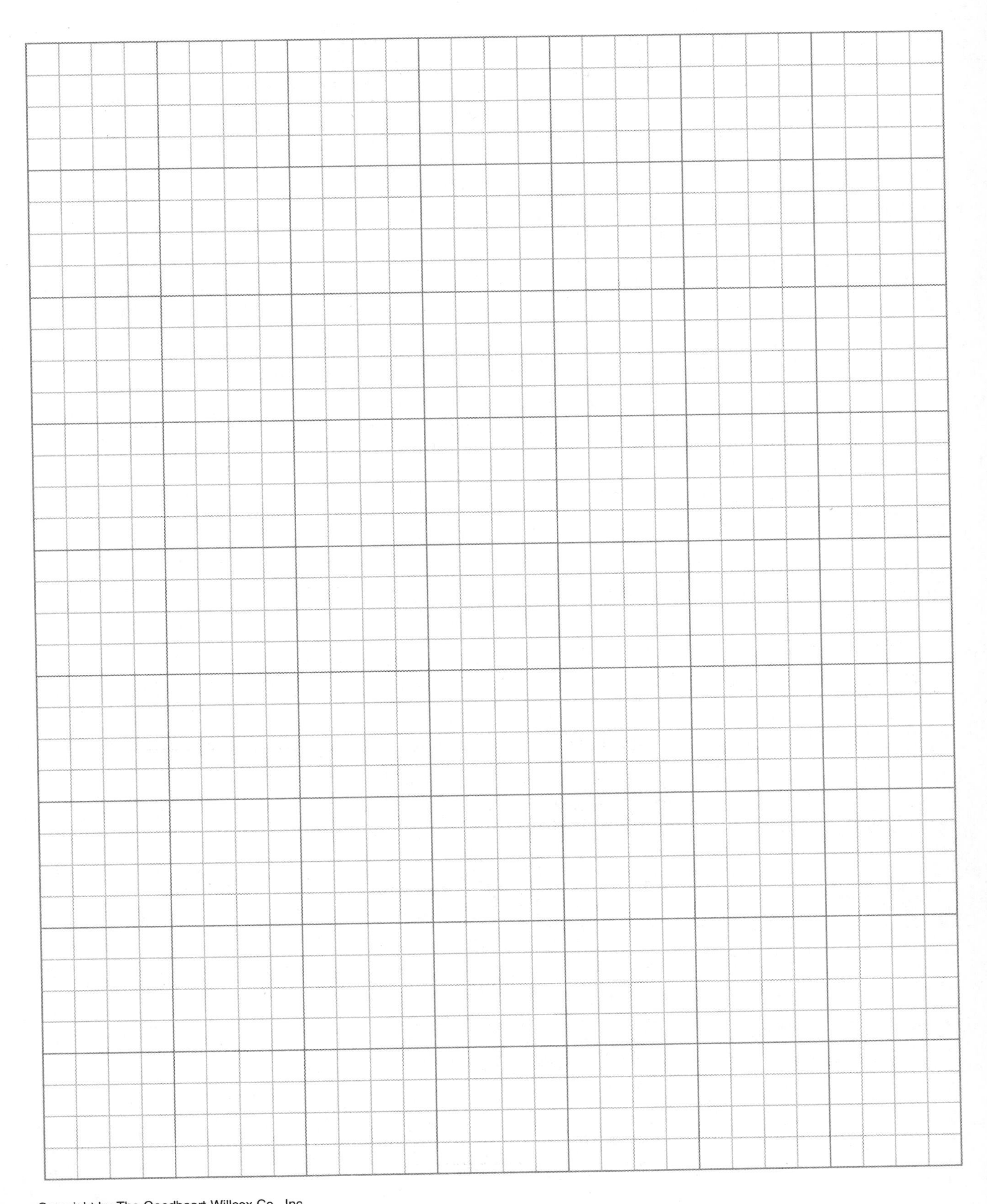

Additional graph paper

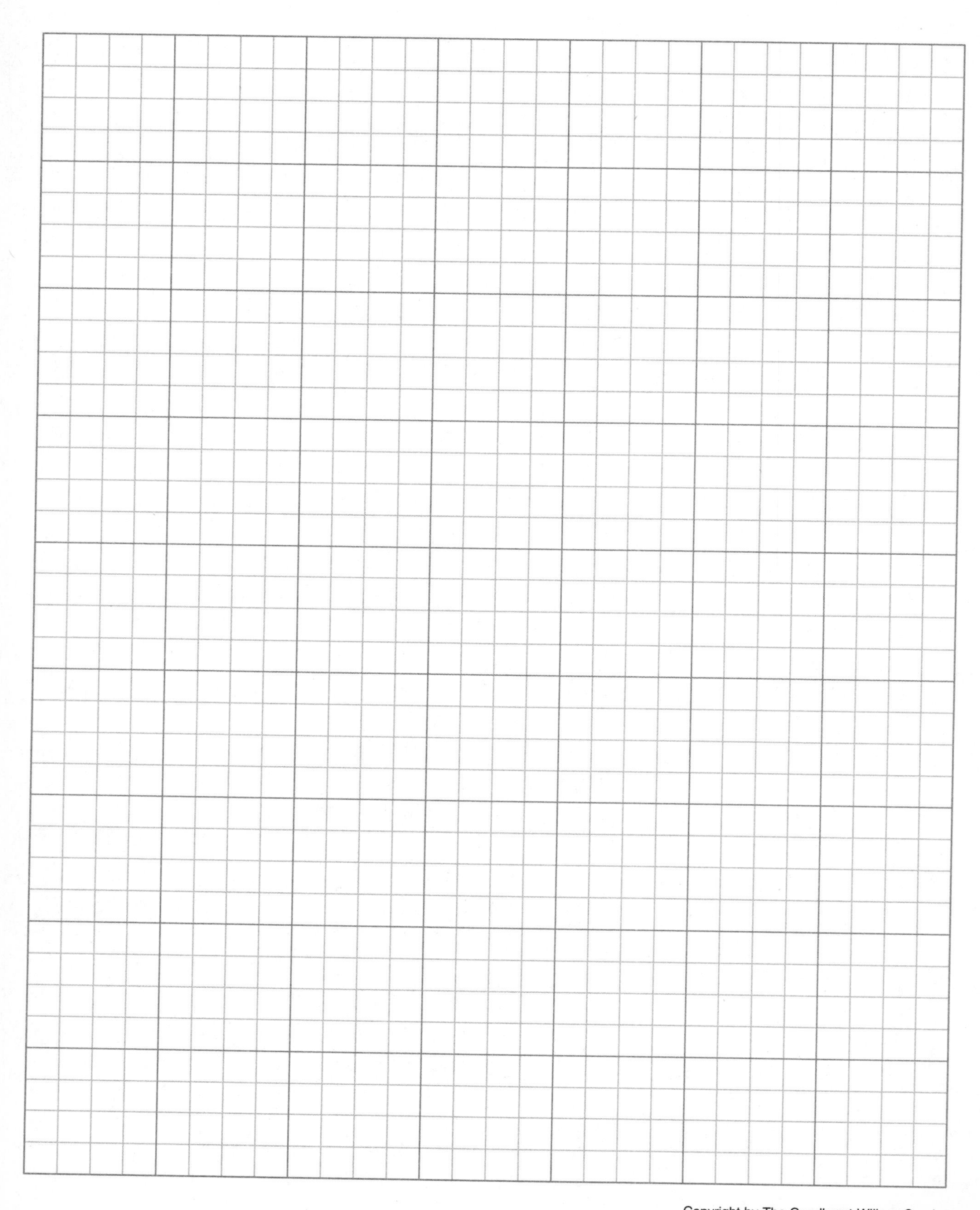

Chapter 15 Review
Chemical Engineering

Name ______________________ Date __________ Class __________

1. Give two differences between the jobs of chemists and chemical engineers.

____

____

____

____

Matching

_______ 2. Entropy approaches zero as temperature approaches absolute zero.

_______ 3. Concentrated energy will naturally disperse, and all matter seeks a state of uniformity.

_______ 4. Energy in an enclosed loop remains constant, and energy cannot be created or destroyed.

_______ 5. When two objects are separately in thermodynamic equilibrium with a third object, they are in thermodynamic equilibrium with each other.

A. First law of thermodynamics

B. Second law of thermodynamics

C. Third law of thermodynamics

D. Zeroth law

_______ 6. _____ describes the fact that the amount of material going into and coming out of a reaction are the same.

A. Entropy

B. Zeroth law

C. Mass balance

D. Fluid dynamics

_______ 7. ____ flow is smooth flow with no disruptions or eddies.

A. Turbulent

B. Laminar

C. Even

D. Fluid

Matching

_______ 8. Flowmeters that measure the flow in a channel rather than in a pipe.

_______ 9. Flowmeters that measure the speed of passing fluid.

_______ 10. Flowmeters that rely on Bernoulli's principle.

A. Open-channel

B. Differential pressure

C. Velocity

Matching

_______ 11. Measure pressure using fluid in a tube.

_______ 12. Measure temperature based on expansion and contraction of unlike metals.

_______ 13. Pressure gauges that use coils of tubing.

_______ 14. Measure temperature based on voltage created by unlike metals.

A. Thermocouple sensors

B. Bimetallic

C. Liquid column

D. Bourdon style

_______ 15. The most efficient and economical method for large-scale chemical processes is _____.

A. MSDS

B. repeating

C. continuous

D. batch

_______ 16. The method commonly used to produce small quantities of chemicals is called _____.

A. MSDS

B. repeating

C. continuous

D. batch

Chapter 15 Review *(Continued)* Name ____________________

________ 17. Simple diagrams that are used to show if the chemical process of a plant will work are called _____.

A. piping and instrumentation diagrams

B. process flow diagrams

C. block flow diagrams

D. site layouts

____________________ 18. True or False? OSHA's mission is to inspect chemical site layouts.

________ 19. Which of the following is NOT included on a material safety data sheet (MSDS)?

A. Chemical's common and chemical name

B. Hazard warnings

C. Manufacturer's name and address

D. Recommended sale price

20. How can chemical engineers help find ways to make coal a safe, environmentally friendly, clean, abundant energy source?

____

____

____

____

Notes

Name ______________________ Date __________ Class __________

Activity 15-1

Mass Balance

Chemical engineers use mass balance to examine chemical processes. The law of conservation of matter states that matter is not created or destroyed. This is the foundation for mass balance. The mass of chemicals prior to a chemical process is equal to the mass of chemicals after a chemical process. In this activity, you will conduct a mass balance test.

Objectives

After completing this activity, you will be able to:

- Conduct a mass balance test.
- Explain how mass is used to evaluate chemical processes.

Safety

Safety requirements will depend on the nature of the materials used in this activity and will be outlined by your teacher.

Materials

Powdered drink mix

Water

Ice

Containers to hold the ingredients and the final product

Scale

Pencil

Activity 15-1 Worksheet

Activity

In this activity, you will:

1. Perform a mass balance test on a simple mixing operation.
2. Weigh the ingredients prior to their mixing.

3. Weigh the final product.
4. Record the weight of the ingredients on Activity 15-1 Worksheet.

Reflective Questions

1. Did your total input mass equal your total output mass? Why or why not?

____

____

____

____

2. What does this tell you about mass balance?

____

____

____

____

Name ______________________________

Activity 15-1 Worksheet

Measure the weight of the following. Remember to weigh the containers separately and subtract their weight from the material weight.

____________________ Powdered drink mix

____________________ Water

____________________ Ice

____________________ Total input materials

Mix the ingredients together and weigh your output materials.

____________________ Total output

Notes

Name ______________________________ Date ____________ Class __________

Activity 15-2

Viscosity

Viscosity is the thickness of a liquid, or its resistance to being deformed. Viscosity is extremely important in chemical engineering because it has a profound effect on chemical processes. In this activity, you will measure the gallons per minute of at least two liquids with different viscosity.

Objective

After completing this activity, you will be able to:

- Perform a viscosity test comparing various materials.

Safety

Safety requirements will depend on the nature of the materials used in this activity and will be outlined by your teacher. Be sure to take the proper precautions when working with chemicals.

Materials

1 quart of motor oil (or similar thick liquid)

1 quart of water

1 funnel with a very small hole

Container to catch the water

Container to catch the oil

Stopwatch or timer

Heat source (optional)

Pencil

Activity 15-2 Worksheet

Activity

In this activity, you will:

1. Pour the water into the funnel and time how long it takes for all of the water to run through the funnel. Record your results on Activity 15-2 Worksheet.
2. Because there are four quarts in a gallon, you can multiply your time by four to find how long it would take for one gallon of water to flow through your funnel. Record your results on Activity 15-2 Worksheet.
3. Repeat this process using motor oil or other thick liquid. Record your results on Activity 15-2 Worksheet.
4. As an optional step, warm the oil, but not to the point where it could cause injury. Repeat the viscosity test with the warm oil.

Reflective Questions

1. What did you observe about the viscosity of oil compared to water?

____

____

____

____

2. Did heating the oil change its viscosity?

____

____

____

Name ______________________________

Activity 15-2 Worksheet

Water:

______________ Total time for one quart of water to flow through the funnel.

______________ Time it would take for one gallon of water to flow through the funnel.

______________ Calculate gallons per hour of flow through the funnel.

Motor oil or other thick liquid:

______________ Total time for one quart of oil to flow through the funnel.

______________ Time it would take for one gallon of oil to flow through the funnel.

______________ Calculate gallons per hour of flow through the funnel.

Warmed motor oil or other thick liquid:

______________ Total time for one quart of warm oil to flow through the funnel.

______________ Time it would take for one gallon of warm oil to flow through the funnel.

______________ Calculate gallons per hour of flow through the funnel.

Notes

Name ______________________________ Date ____________ Class __________

Activity 15-3

Flow Test

Chemical engineers must be able to design, measure, and monitor flow rate for chemical operations. Fluid flow rate can be measured using a variety of devices. In this activity, you will calculate flow rate by finding out long it takes for a faucet to fill a one-gallon container.

Objective

After completing this activity, you will be able to:

- Use the bucket and timer to determine gallons per minute of fluid flow.

Safety

Follow all safety procedures outlined by your teacher.

Materials

One-gallon container or other reservoir with gallons marked on it

Stopwatch or timer

Faucet and sink large enough to fit a one-gallon container

Pencil

Activity 15-3 Worksheet

Activity

In this activity, you will:

1. Find the flow rate of a faucet.
2. Record your results on Activity 15-3 Worksheet.

Reflective Questions

1. Where might this method of flow measurement be useful?

2. How could you change this experiment to make it more accurate?

Name ______________________________

Activity 15-3 Worksheet

Time how long it takes for a faucet to fill a one-gallon container.

______________________ Total time it took to fill the one-gallon container

You can then use the findings to calculate the gallons per hour of water flow from the faucet. Use the space below for your calculations.

______________________ Gallons per hour flow rate

Notes

Name _________________________________ Date ____________ Class __________

Activity 15-4

Chemical Plant Design—Block Flow Diagram

In this activity, you will draw a block flow diagram for a possible chemical plant process.

Objective

After completing this activity, you will be able to:

- Draw a block flow diagram for a chemical plant process.

Materials

Pencil

Activity 15-4 Worksheet

Activity

In this activity, you will:

1. Use Activity 15-4 Worksheet to draw a block diagram for three feed chemicals that are heated, mixed, and cooled.

Reflective Question

1. For what product(s) can you imagine this process being used?

____

____

____

____

____

Activity 15-4 Worksheet

In the space provided below, draw a block flow diagram for three feed chemicals that are heated, mixed, and cooled. The chemicals start out in separate containment vessels. Each then moves to a heating process and then into a mixer. After the chemicals are mixed, they move to a cooling process and then to a storage vessel.

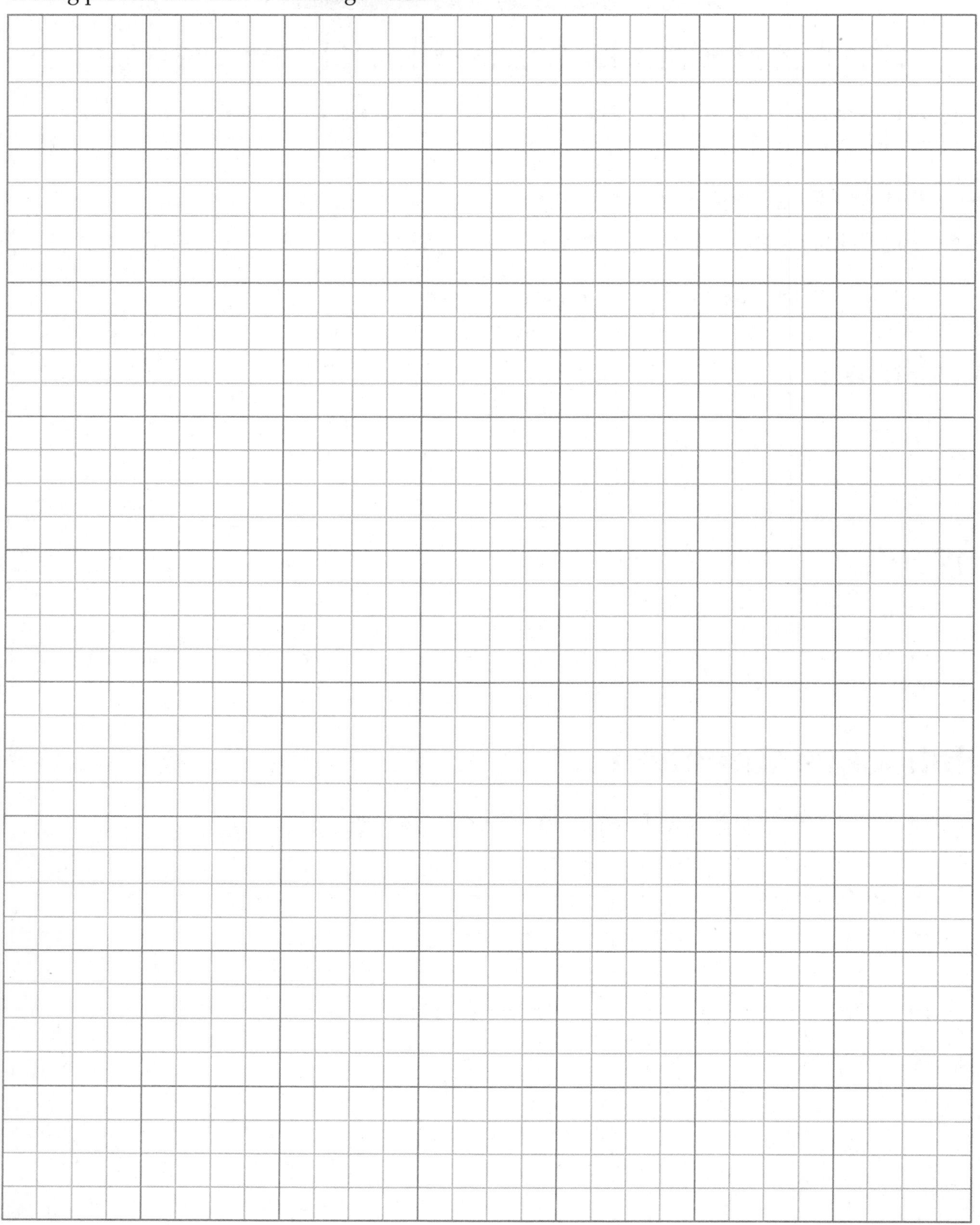

Name ______________________________ Date ______________ Class ____________

Activity 15-5

Chemical Plant Design—Site Layout

Objective

After completing this activity, you will be able to:

- Draw a simple sketch of a chemical plant site layout.

Materials

Pencil

Activity 15-5 Worksheet

Activity

In this activity, you will:

1. Use Activity 15-5 Worksheet to design and draw a simple sketch of a chemical plant site layout.
2. Include the components listed on Activity 15-5 Worksheet.

Reflective Question

1. How is the site layout different from the block flow diagram from the previous activity?

____

____

Activity 15-5 Worksheet

The design should plan for the following:

- Storage of feed chemicals
- Storage of finished chemical products
- Convenient traffic flow for cars and large trucks
- Utilities (electricity and natural gas)
- Maintenance department facilities
- Possible future expansion
- Storm water management

Activity 15-5 Worksheet *(Continued)* Name ______________________

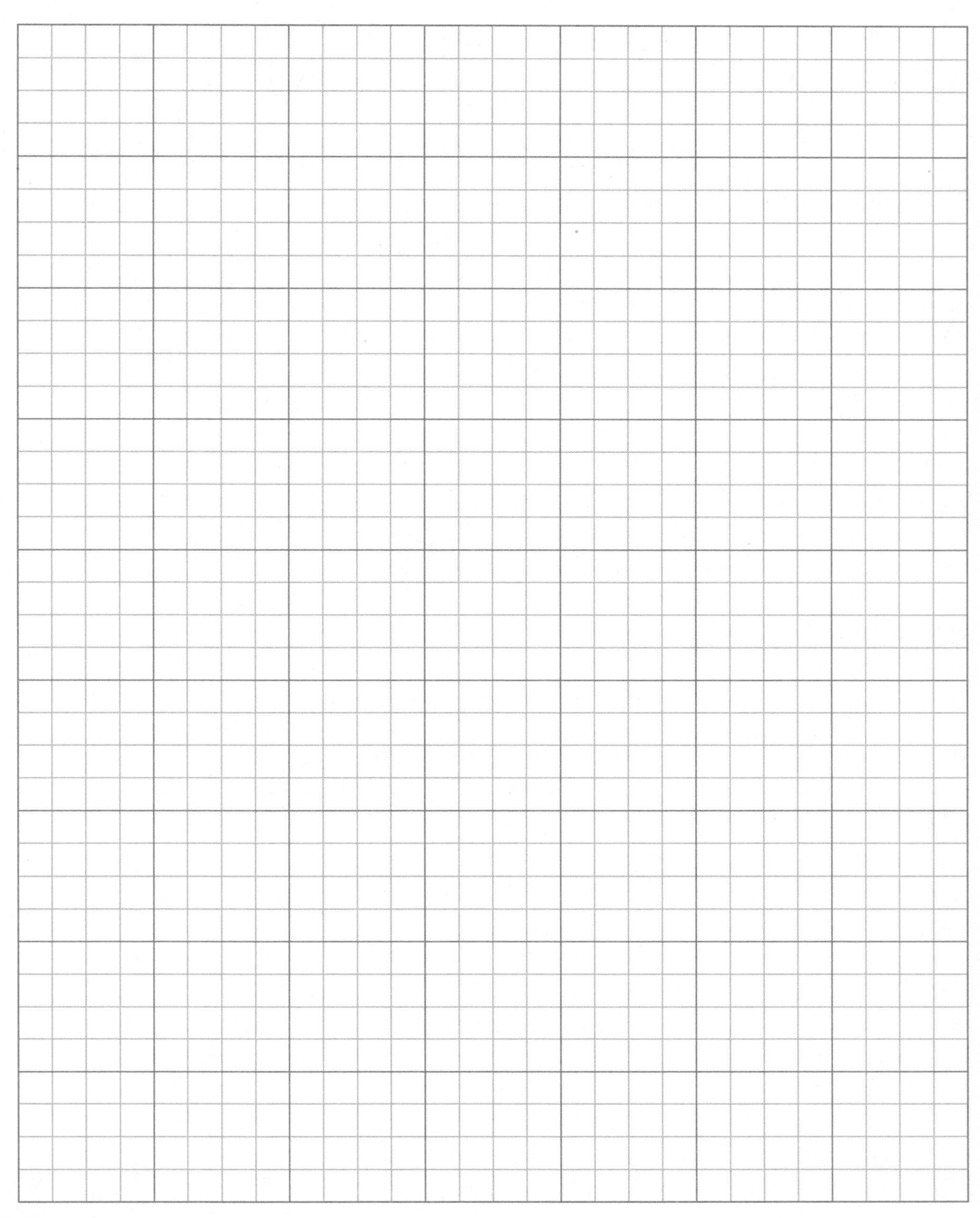

Additional sketching paper

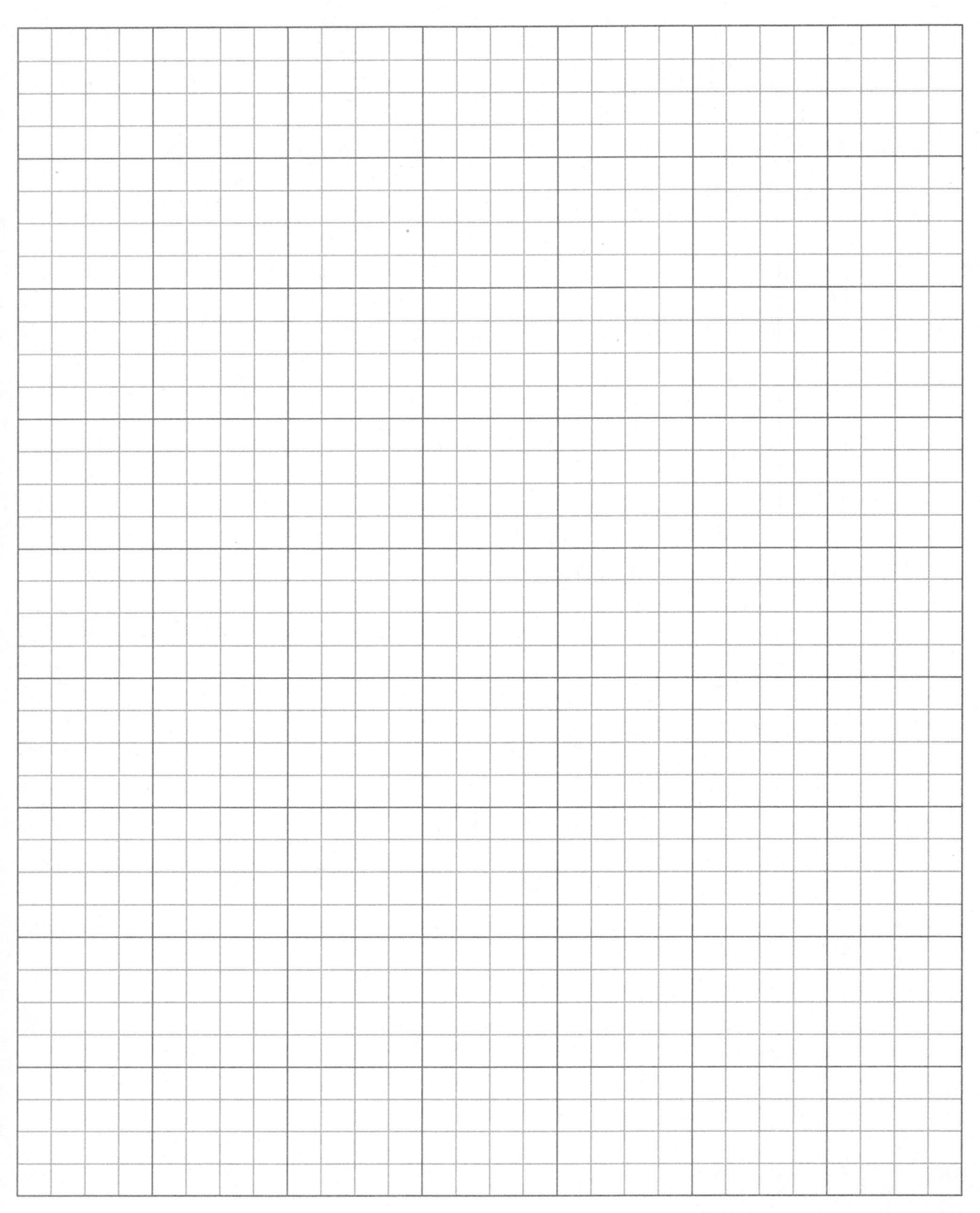

Name ____ Date ____________ Class __________

Activity 15-6

Material Safety Data Sheets

Material safety data sheets (MSDS) list the chemical's common and chemical name, hazard warnings, first aid for exposure, disposal information, transportation requirements, and the manufacturer's name and address. In this activity, you will find and read MSDS for everyday products and understand the information.

Objective

After completing this activity, you will be able to:

- Read and understand material safety data sheets.

Materials

Sample MSDS (easily found for free online)

Computer with Internet access

Pencil

Activity 15-6 Worksheet

Activity

In this activity, you will:

1. Look up an MSDS for a material you might encounter in your daily life.
2. Read the document and answer the questions on Activity 15-6 Worksheet.

Reflective Questions

1. Did you learn something about a common product that you did not know?

    ____

2. Why do you think having the information on an MSDS is important to people who work with the product?

    ____

    ____

Activity 15-6 Worksheet

1. How and where might you come in contact with this material?

2. Should you seek medical attention if you come in contact with this material?

3. What should people do if they get this material in their eyes?

4. If this material is ingested, should you induce vomiting?

5. Is this material flammable?

Chapter 16 Review

Engineering as a Profession

Name ______________________________ Date ____________ Class ___________

Matching

_______ 1. Research engineers

_______ 2. Development engineers

_______ 3. Operations engineers

A. Engineers who work with scientist to find uses of scientific discoveries.

B. Engineers who control and maintain large systems.

C. Engineers who test and analyze ideas and products.

4. Why are communication skills important for engineers?

____

____

5. What type of degree would you need to become an engineering technician?

____

6. High-quality engineering programs are accredited by _____.

____

_________________ 7. True or False? All engineers must obtain the Professional Engineer (P.E.) license.

8. What is an example of an ethical principle in the engineering profession?

____

____

9. What is the main priority in engineering ethics?

____

10. List two examples of societal impacts of engineering.

____

____

11. What is the most immediate economic impact from the work of engineers?

12. Explain the purpose of a patent.

13. What are two movements that stress sustainability and positive environmental impacts?

14. Provide two examples of the positive impacts (benefits) of engineering.

15. Provide two examples of the negative impacts (costs) of engineering.

Name ______________________________ Date ____________ Class __________

Activity 16-1

Engineering Roles

Engineers perform a number of functions within the engineering profession. In this activity, you will select an engineered product and determine functions engineers may have performed on the product.

Objective

After completing this activity, you will be able to:

- Provide examples of engineering functions.

Materials

Pencil

Activity 16-1 Worksheet

Activity

In this activity, you will:

1. Select an engineered product, such as a bridge, prosthetic, or electrical device.
2. Imagine the process the engineers went through to design the selected product.
3. Using Activity 16-1 Worksheet, list tasks that would have been performed throughout the process in each of the functions.
4. Present your findings to the rest of the class.

Reflective Questions

1. Is there a function of engineering that is interesting to you as a profession?

____

____

2. Are there functions of engineering that you did not know existed?

____

____

Activity 16-1 Worksheet

Function	Tasks
Design	
Research	
Development	
Production	
Sales	
Management	

Name ______________________ Date __________ Class __________

Activity 16-2

Engineering Impacts

Engineering products, processes, and systems have a wide range of impacts. The work of engineers impacts several areas of our lives and can be both positive and negative. In this activity, you will research and present on the impacts of an engineered product.

Objective

After completing this activity, you will be able to:

- Describe impacts of engineering.

Materials

Pencil

Activity 16-2 Worksheet

Presentation software

Activity

In this activity you will:

1. Select a engineered product, such as a bridge, prosthetic, or electrical device.
2. Conduct research on the engineering product to gain an understanding of its impacts.
3. Complete the table in Activity 16-2 Worksheet as you identify the areas and types of impacts of the product.
4. Using presentation software, create a presentation to display your findings.
5. Present your findings to the rest of the class.

Reflective Questions

1. Do you believe that all of the benefits and costs were known during the design of the product?

____

____

2. How do engineers weight the benefits and costs and make decisions?

____

____

Activity 16-2 Worksheet

Product:	Tasks of Impacts	
	Positive Impacts (Benefits)	**Negative Impacts (Costs)**
Personal Impacts		
Societal Impacts		
Economic Impacts		
Environmental Impacts		